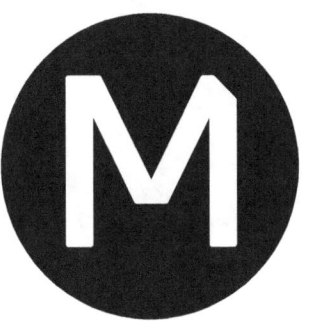

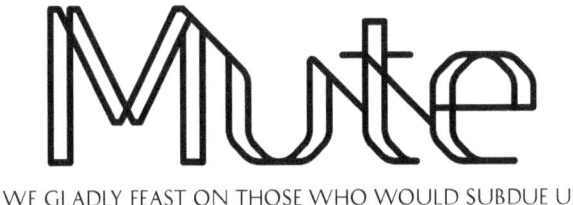

WE GLADLY FEAST ON THOSE WHO WOULD SUBDUE US

VOLUME 3 #2
POLITICS MY ARSE
AUTUMN/WINTER 2011-12

Published by Mute Publishing
ISBN 978-1-906496-02-9
ISSN 1356-7748-302

Also available as eBook ISBN 978-1-906496-77-7

In collaboration with Post-Media Lab, Leuphana University

Mute Publishing is supported by
Arts Council England

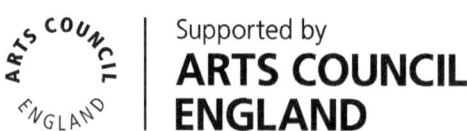

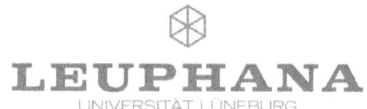

MUTE VOL 3 #2
AUTUMN/WINTER 2011-12

EDITOR
Josephine Berry Slater
<josie@metamute.org>

EDITORIAL BOARD
Josephine Berry Slater

Omar El-Khairy
<omarelkhairy@gmail.com>

Matthew Hyland

Anthony Iles
<anthony@metamute.org>

Demetra Kotouza
<demetra@inventati.org>

Hari Kunzru
<hari@metamute.org>

Mira Mattar
<miramattar@googlemail.com>

Pauline van Mourik Broekman

Benedict Seymour
<ben@metamute.org>

Stefan Szczelkun
<stefan@szczels.plus.com>

Simon Worthington

MUTE PUBLISHING ADVISORY BOARD
Sally Jane Norman

Sukhdev Sandhu

Andrew Seto

Andrew Wilson

PUBLISHERS
Pauline van Mourik Broekman
<pauline@metamute.org>

Simon Worthington
<simon@metamute.org>

ISSUE DESIGN
Atwork
http://www.atworkportfolio.co.uk

ADVERTISING & MARKETING
T: +44 (0)20 3287 9005
E: <mute@metamute.org>

WEBSITE
Metamute.org is run on Drupal FLOSS Software, with additional software services by our very own OpenMute http://openmute.org. Graphic design by Atwork, CSS by Roglok <roglok@hyperground.de>, template coding by Effusion http://effusion.co.uk

CHIEF ENGINEER
Darron Broad <darron@kewl.org>

LAYOUT DESIGN
Laura Oldenbourg <laura@metamute.org>
Raquel Perez de Eulate <raquelwebs@googlemail.com>

PROJECT CO-ORDINATOR / OFFICE MANAGER
Caroline Heron <caroline@metamute.org>

INTERNS
Marisa Boyle & Mira Mattar

IMAGE REPROGRAPHICS
Happy Retouching <richard@happyretouching.com>

OFFICE
Mute, 46 Lexington Street, London, W1F 0LP
T: +44 (0)20 3287 9005
E: <mute@metamute.org>

SUBSCRIPTIONS
Howard Slater
T: +44 (0)20 3287 9005
E: <howard@metamute.org>
W: http://www.metamute.org/subs

DISTRIBUTION UK
Central Books, 99 Wallis Road, London E4 5LN
T: +44 (0)20 8986 4854
F: +44 (0)20 8533 5821
E: <mark@centralbooks.com>

CONTRIBUTING
Mute welcomes contributions of all kinds. Email <mute@metamute.org> with your ideas. You can also publish on Mute's website [metamute.org]. Post news, text, events and comments, or upload media. The views expressed in Mute and Metamute are not necessarily those of the publishers or service providers. Mute is published in the UK by Mute publishing Ltd. and printed by OpenMute [http://openmute.org] print on demand [POD] book services.

COVER
Johnny Spencer <mail@johnnyspencer.info>
http://www.johnnyspencer.info

SPECIAL THANKS
Thanks to Paul Graham and Ian Rich for invaluable proofing work.

DORTMUNDER U
ZENTRUM FÜR KUNST
UND KREATIVITÄT

HMKV
Hartware MedienKunstVerein

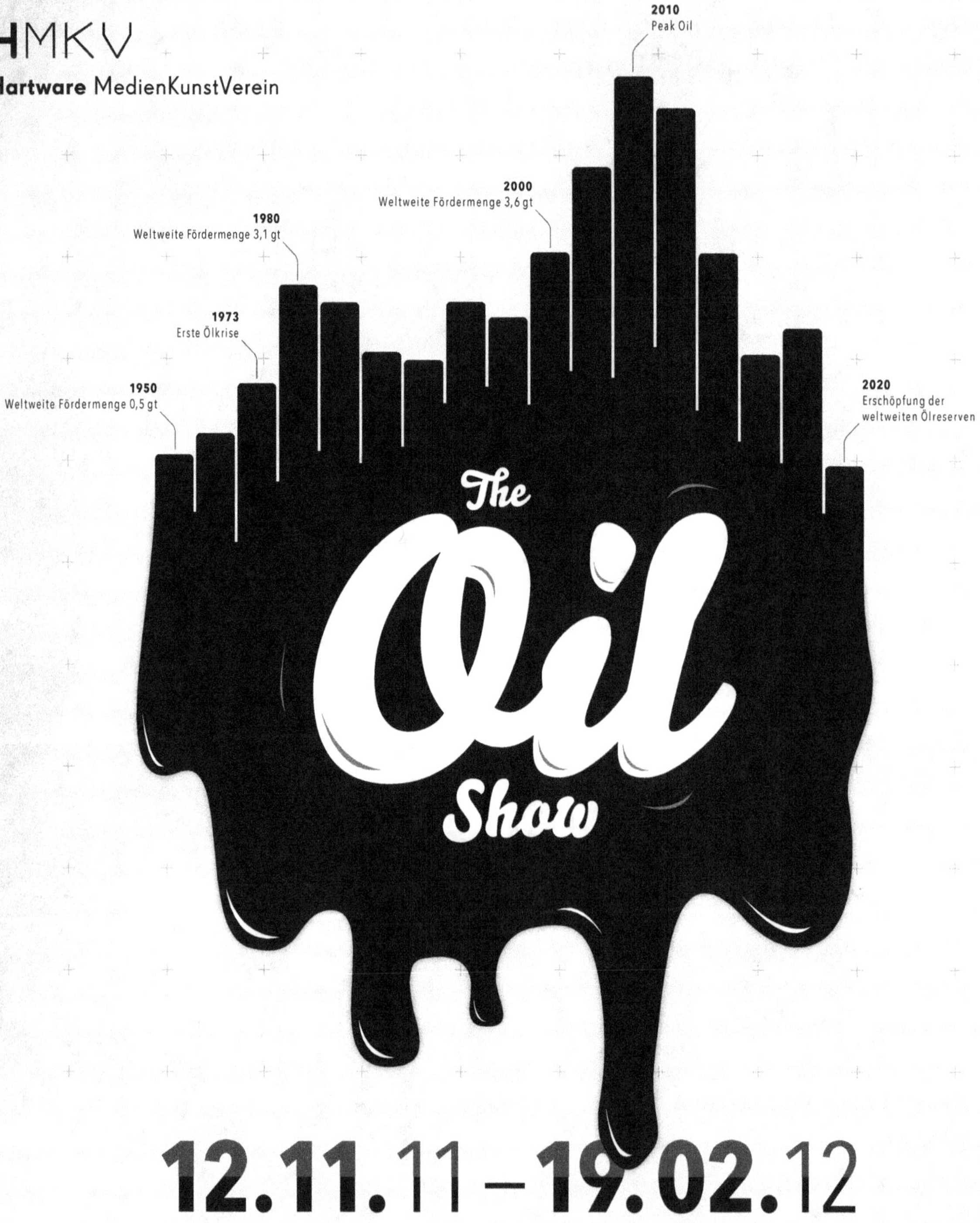

2010
Peak Oil

2000
Weltweite Fördermenge 3,6 gt

1980
Weltweite Fördermenge 3,1 gt

1973
Erste Ölkrise

1950
Weltweite Fördermenge 0,5 gt

2020
Erschöpfung der
weltweiten Ölreserven

The
Oil
Show

12.11.11 – 19.02.12

HMKV IM DORTMUNDER U **WWW.HMKV.DE**

WWW.DORTMUNDER-U.DE

PARTNER 2011   HAUPTFÖRDERER DES HMKV

MIT FREUNDLICHER UNTERSTÜTZUNG DER   MEDIENPARTNER

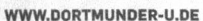   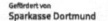

KUNSTSTIFTUNG ● NRW   Ministerium für Familie, Kinder, Jugend, Kultur und Sport des Landes Nordrhein-Westfalen   Gefördert von Sparkasse Dortmund   DSW21   schweizer kulturstiftung prohelvetia   Bürgschaft der Künstlerin   **arte** CREATIVE ●DE

Afterall
28

Autumn/Winter 2011

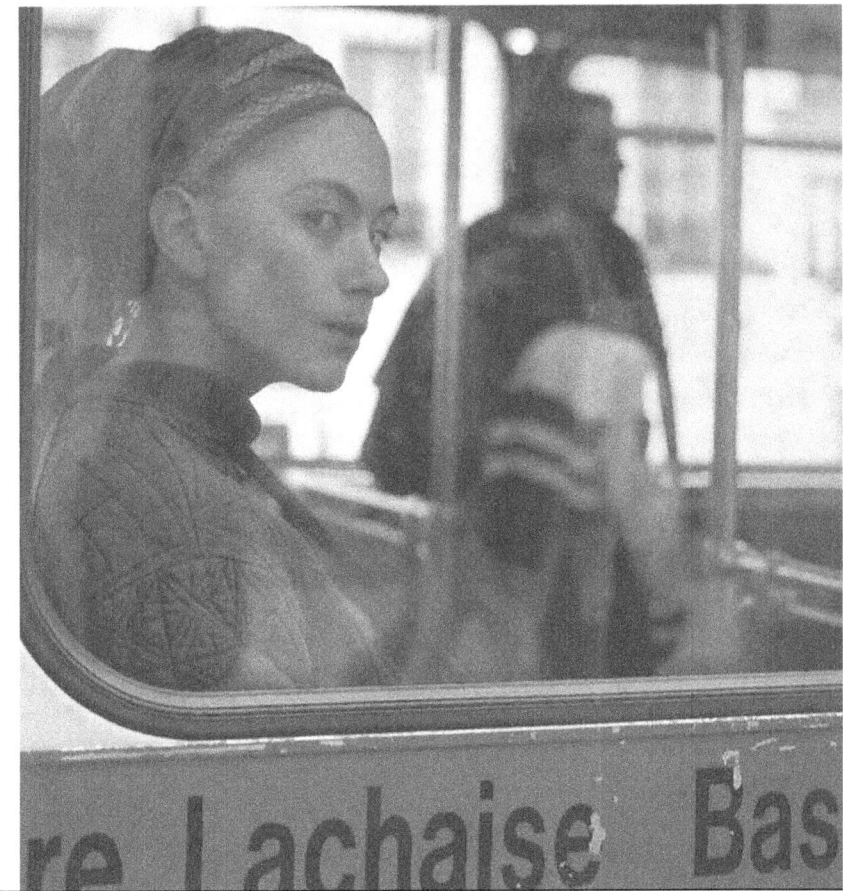

Afterall
One Work

Autumn/Winter 2011

Jeff Koons

One Ball Total Equilibrium Tank

Michael Archer

Afterall Books

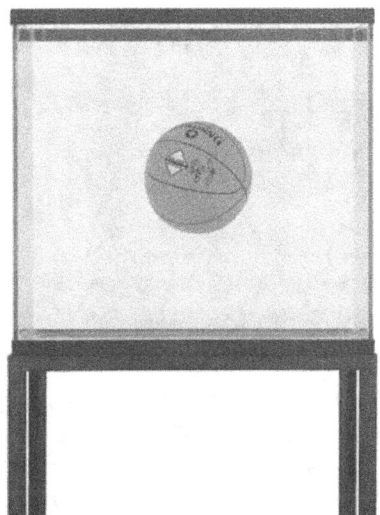

Afterall Books are distributed by The MIT Press
http://mitpress.mit.edu/afterall

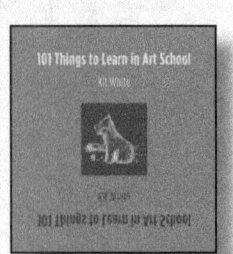 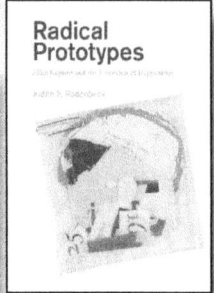 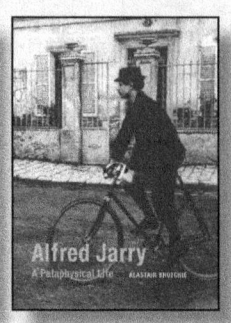 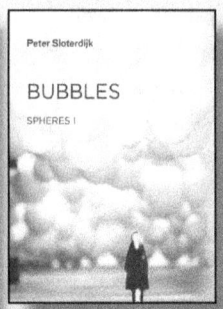 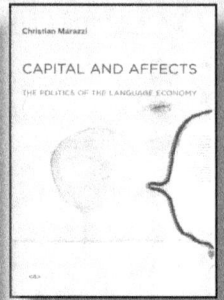

101 THINGS TO LEARN IN ART SCHOOL
KIT WHITE

Lessons, demonstrations and tips on what to expect in art school, what it means to make art and how to think like an artist.

£10.95 • 224 pp. (101 illus.) • cloth • 978-0-262-01621-6

RADICAL PROTOTYPES
Allan Kaprow and the Invention of Happenings
JUDITH F. RODENBECK

An examination of an experiential and experimental art form that, despite its evanescence, has shaped participatory art into the present.

£24.95 • 312 pp. (47 illus.) • cloth • 978-0-262-01620-9

ALFRED JARRY
A Pataphysical Life
ALASTAIR BROTCHIE

"It is rich in fresh research and archival treasures ... Brotchie's painstaking and drily funny biography is now the most ample account of Jarry and his importance that is available in our language; it is unlikely to be bettered."
– **Kevin Jackson**, *The Literary Review*

£24.95 • 424 pp. (156 illus.) • cloth • 978-0-262-01619-3

BUBBLES
Spheres I: Microspherology
PETER SLOTERDIJK
translated by Wieland Hoban

Rejecting the century's predominant philosophical focus on temporality, Sloterdijk reinterprets the history of Western metaphysics as an inherently spatial and immunological project, from the discovery of self (bubble) to the explanation of world (globe) to the poetics of plurality (foam). Exploring macro- and micro-space from the Greek agora to the contemporary urban apartment, Sloterdijk is able to synthesize, with immense erudition, the spatial theories of Aristotle, René Descartes, Gaston Bachelard, Walter Bejamin, and Georges Bataille.

£24.95 • 664 pp. (143 illus.) • cloth • 978-1-58435-104-7 • *Distributed for Semiotext(e)*

CAPITAL AND AFFECTS
The Politics of the Language Economy
CHRISTIAN MARAZZI
translated by Giuseppina Mecchia

Christian Marazzi's first book, now a foundational work in post-Fordist literature, goes beyond political economy to encompass issues related to social life, political engagement, democratic insitutions, interpersonal relations and the role of language.

£10.95 • paper • 160 pp. • 978-1-58435-103-0 • *Distributed for Semiotext(e)*

http://mitpress.mit.edu
orders: 01243 779 777

The MIT Press

THE SHOW ROOM

63 Penfold Street
London NW8 8PQ
T/F: 020 7724 4300
www.theshowroom.org

SIGNAL:NOISE, PART 2

Friday 20 – Saturday 21 January 2012

Building on the success of Signal: Noise, Part One in January 2011, this two day event continues research into the influence of cybernetics and information theory on contemporary cultural life by testing out its central idiom 'feedback' through debates, performances, and events. Organised in collaboration with Mute and the School of Business and Management, Queen Mary, University of London.

for more information visit www.theshowroom.org

The Showroom is supported by Arts Council England and members of the gallery's Supporters Scheme.
Signal:Noise, Part Two is supported by LCACE and Queen Mary, University of London

VOLUME 3 #2
POLITICS MY ARSE
AUTUMN/WINTER 2011-12

EDITORIAL

In more 'ordinary' times, the sense of all those unwritten articles we'd like to commission for *Mute* can be quite overwhelming. Just working through the stack of review copies and printed matter that threaten to bury us alive in our office is a sizeable enough task. Then, looking up from the stack, there's the imposing rock face of human activity in general, and its often inhuman consequences, to be considered. Which pathways should be sheered into it, which abandoned? But these days, the need for coverage has gone into warp drive, as the stable coordinates of our existence are thrown in the air. How can we contend with a picture as complex and fast moving as that of our present moment?

Last year, as Sarah Taylor reminds us [p.14], the prospect of FE and higher education becoming a blatant privilege of the well off seemed, well, scandalous – if only for its official acknowledgement of what we already knew to be the case. But this pales by comparison with the vertical drop in basic living standards now being imposed across Europe, or the casual removal of the last veil of democracy – namely elected politicians – in favour of technocrats who actually 'know what they're doing'. The way that the founding principles of 'decent' society (read, remnants of social democracy) become, in a heartbeat, the unaffordable luxuries of yesterday, creates a vertiginous sense of how fast and how far this crisis is taking us. The vertigo is light and heavy at the same time.

In such end times, where answers seem unlikely, even suspect, the question of 'what is to be done?' haunts us everywhere. A friend recently advised reading Robert Musil's *The Man Without Qualities* for its sharp delineation of a collective desire to act, the pursuit of which becomes a collective pseudo-activity in turn, preventing anything from *actually* happening. One could say, by extension, that publishing and consuming articles that inform, comment on and diagnose our shattering social relations seems critical right now, and yet the production and proliferation of text can also feel like yet another pseudo-activity blocking us from actually *doing* anything.

It's tempting to imagine a situation like that of Egypt last January when the authorities switched off the internet, compelling thousands to flock to Tahrir Square in order to find out what on earth was going on, inadvertently triggering the critical mass with which the 'revolution' was carried through. Of course we have to balance this picture out against last August's 'Blackberry riots' in which, or so the moral panic asserts, it was precisely the circulation of information, BBM, SMS and social media, which created mass insubordination and unpredictable swarm effects. The 'text-act' becomes what Howard Slater discusses in these pages as the 'language of acts' which resolve, in turn, into more mass messaging [p.74]; a feedback loop of cascading effects.

The anti-capitalist engagement with the global conduits and logistics of circulation and communication presents a related dilemma of perspective, of the too zoomed out and the too close in. Arguing against the anarchistic desire to induce system collapse by sabotaging these circuits, Alberto Toscano [p.30] considers how the urban concentrations produced by capitalism have reconfigured human life on a scale that cannot simply be wished away by those looking for new, unalienated realities. In short, we need logistics and capitalism's advanced means of production and circulation in order to survive as a largely urban species, but how can we repurpose them? What would communist logistics look like?

These sketches pose the same question of how we engage with disaster capitalism from multiple scales, holding the abstract and the first person perspectives together at once. If we spend too long becoming fetishists of a systems analysis, embellishing diagrams – whether textual, informational or discursive – of our own oppression, we risk becoming mesmerised by their complexity. On the other hand, going down into the streets, speaking face to face with our comrades and cutting the power lines which entrap (and produce) us, doesn't guarantee the Tahrir Effect either. And as we know, the feel good factor of squares can detract from the painful work of dismantling and rebuilding societies so as not to leave the bad 'old minds', as an Egyptian activist had it, at the heart of the system.

What we need is some kind of alternating current which can bring situations of immediacy as well as representations adequate to the moment's complexity into relation. This idea brings to mind a recent project by YoHa, an art group featured in the last issue of *Mute*, in which Bristol council's newly open databases were harvested to build contraptions which both reveal and convert the data's function. In one case, residents could engage in the 'book stabbing' of budgetary data or, in another, riding a bicycle seat connected to the indices of council expenditure as it rises and falls. This provided both a venting of unconscious sadism stored up by our subjection to municipal technocracy as well as an enlightened understanding of how data-sets are the endlessly divertable servants (and drivers) of the systems they dwell within.

YoHa's jamming together of data with absurdist and gory actions, bringing representation and its material effects into new proximities, points in a liberating direction. Displacing representations, analyses and data-sets of oppressive social relations so that, at the very least, they are unable to perpetuate an image of business as usual, is a way to use our scant resources potently. A residing image I have of visiting the occupation at St. Paul's is of a woman sticking up printouts from WikiLeaks with hazard tape. This neat unfolding of impacted governmental data into the pseudo-public arcade at the edge of Paternoster Square invites a new way of reading which quite literally borders on action.

JOSEPHINE BERRY SLATER
<josie@metamute.org> is Editor of *Mute*

WHAT NEXT FOR EDUCATION STRUGGLES?

After last August's riots SARAH TAYLOR *asks, will the contradictions of last year's student movement resolve or simply extend themselves?*

Last year was so simple. The government wanted to triple fees and there was a vote coming up. Smashing Millbank made everything seem possible. Demonstrations stopped going where they were told, school students played truant to run riot in Oxford Street, occupations at universities became the norm. But the vote was lost. Tired and cold, we went home and didn't come back.

A year later, the lectures and lessons have restarted and the student demonstrations are planned again. The first is on 9 November[*] – a year since the occupation of Millbank. But what do we do on this anniversary? If it was all about the vote on tuition fees and keeping Educational Maintenance Allowance, then perhaps we should lay a wreath for the lost fight – for the university students filming themselves Twittering in occupations, and for the school students smashing up shops, for dance-offs in libraries and grime in kettles.[1] But if it was about something more, and if all these people are still angry, what happens next?

National Campaign Against Fees and Cuts (NCAFC) and the Education Activist Network (EAN) called the 9 November protest. The National Union of Students (NUS), now apparently resigned to its insignificance, officially supports the demonstration, and has agreed to provide some resources to help publicise it, although this doesn't seem to stretch to the use of its website, which currently headlines 'Freshers' advice from the cast of *Hollyoaks*.'[2] Filling the hole left by the NUS, the new student organisations are looking a lot like the old left. EAN was never shy of being the offspring of the ageing Socialist Workers Party, and although NCAFC was originally a loose

network of students, over the summer it elected itself a National Committee with 14 permanent members and now describes itself as one of the organisations that 'led' the student protests.[3]

It works both ways – the old left also wants to emulate the students. Last year, Unite's Len McCluskey wrote in *The Guardian*,

> Britain's students have certainly put the trade union movement on the spot. Their mass protests against the tuition fees increase have refreshed the political parts a hundred debates, conferences and resolutions could not reach [...] The magnificent students' movement urgently needs to find a wider echo if the government is to be stopped.

But he also thinks that any 'wider echo' will need trade unions to give 'guidance' to people's anger, to 'put it in a manner that will hopefully make the government take a step back.'

On 30 November, public sector workers will walk out over pensions, in what is set to be the biggest strike since the 1926 general strike. Education workers are the most obvious point of contact between the strikers and the students. Most teaching unions, ranging from the large and relatively militant National Union of Teachers to the small and conservative Association of Teachers and Lecturers, will be walking out. The higher and further education union, the University and College Union, will also be on strike. In a show of solidarity, the NUS leader warns, 'Any action that threatens students' ability to progress from year to year, or graduate at all, will immediately lose student support.'[4] Leaving the NUS to the cast

* This article was written before the 9 November sudent demonstration and 30 November public sector walkout

of *Hollyoaks*, NCAFC and EAN have called for student walkouts and direct action on the day of the strike. A similar call-out for the smaller 30 June public sector strike met with little response, but away from end of term lethargy, and with more workers on strike, young people might be tempted onto the streets.

Yet, as the last strike indicated, the government is not really fazed by one-day strikes – far from stepping back, it steps over them and keeps on going. Even if students do decide to join in the struggle for pensions, the government is unlikely to feel threatened by 12 hours of adults and children fighting to retire.

But you would be hard pushed to find a subject further from the minds of young people than pensions. Even the striking teachers will tell you that it is work and not retirement that really frightens them. The schools White Paper and the higher education White Paper map out similar plans to deepen the privatisation of the education system: both drive educational institutions into the hands of corporations; both see students and parents as consumers looking for 'choice' or 'value for money'; both threaten teachers' and lecturers' nationally agreed pay and conditions; and both create a two-tier system in which adequately funded education is reserved for the well informed and well moneyed.

The government has engineered the cap on university places in such a way that the Russell Group universities are allowed to siphon off more students with grades AAB and above, and other universities, having lost these students, are encouraged to drop fees below £7,500 to compete with private providers for a pool of 20,000 textra places. Meanwhile, the government's 'free schools' allow groups of middle class parents to open schools, with education corporations taking over when

they find they have bitten off more than they can chew. If that wasn't enough, Labour's Academies, increasingly part of corporation or church-run chains, have been extended to include primary schools. Senior civil servants (who, incidentally, and confusingly, might also join the strike on 30 November) showed their sense of humour by naming the schools White Paper, 'The Importance of Teaching' and the higher education White Paper, 'Students at the Heart of the System'.

But schools are not only subject to direct attacks – they also suffer from the removal of benefits, wages, houses and every other meagre compensation that was previously offered to children and their families. I will take a primary school and secondary school I know as examples of this, but these are by no means isolated cases – the same stories can be heard in schools across the country.

The primary school, where 48 percent of the children are on free school meals, but which is located in a rich London borough, expects to lose half of its pupils once the national housing benefit cap comes into effect – no longer able to afford to rent their own homes, their families will be forced to move from the borough or face eviction. To make things worse, because only three percent of primary schools have taken up Education Minister Michael Gove's generous offer for them to voluntarily become Academies, schools like this one are going to be forced to become Academies next year if they don't fulfil certain, impossible to fulfil, criteria.

The secondary school, based in rural Wales, is facing falling student enrolment as factory after factory in the area is closed and, with dwindling funds from low intake, middle class parents exercising their 'parental choice' decide to send their children elsewhere. This school is likely to be one of half the schools shut down in the county under a PricewaterhouseCoopers consulted 'modernisation' programme, which will see the end of schools that have been the centre of villages and towns for generations, and will force children to travel for up to an hour on country roads before they get to their first lesson.

Once they've left school, more and more of these children won't be able to afford to go to college, let alone university. Many of them won't get jobs, won't get benefits, won't get houses. They can only dream of retirement. The government has something right, students are at the heart of the system, and it beats at them from all sides.

This was clearly expressed in August when, away from the constraints of term time, articulated demands, symbolic targets and organising committees, Britain's teenagers went rioting. The same McCluskey who heaped praise upon the student riots described the August riots as 'the exact opposite of community spirit, collectivism and what trade unionism is all about.'[5] While, a year on from the student riots, many of those arrested are still awaiting trial, the August riots saw courts operating throughout the night, getting people off the streets and into prison as quickly as possible. Some of those arrested during the student protests who were unfortunate enough to be tried immediately after the summer riots, found themselves with harsher sentences than students convicted of exactly the same crimes before the riots. The August riots shook the authorities in a way that the student riots did not. Anything that happens next will have to be seen through the smoke of Tottenham, Croydon, Manchester and Birmingham.

The difference between the winter's student riots and the summer's standard riots is something the student leaders have been

If all these people are still angry, what happens next?

A good riot announces itself with a protest beforehand, has a symbolic target and a view of Big Ben

keen to reinforce. NCAFC patronise the young 'victims', telling them their rioting 'will not improve the situation', and arguing that their anger 'needs to be channelled into tackling the real causes of injustice and inequality.'[6] By which we can only assume they mean Topshop, Camilla Parker-Bowles, an office block in Millbank, the windows of the treasury, Barclays bank, parliamentary votes and the police, rather than Footlocker, the Sony warehouse, JD Sports, Carphone Warehouse, jewellery shops, furniture chains and the police. A good riot announces itself with a protest beforehand, has a symbolic target and a view of Big Ben. A bad riot is localised, unplanned, gets you free trainers, and can even happen when politicians are away on holiday. Good riots have uncontrollable fire extinguishers, bad riots have uncontrollable fires.

The attempt to divide the good and the bad ignores the fact that the summer riots are a continuation of, rather than a break with, the winter riots. The summer riots happened because the winter riots were never going to tackle 'the real causes of justice and inequality.' They happened because the winter riots were survived by a lot of angry people who knew that parliamentary votes had nothing to do with them. They gave up asking for £30 a week – they took what they wanted and destroyed what they hated, and they didn't need to go to central London for that, because it was right outside their doors.

One intrepid reporter of student protest fame spent the August riots 'huddled' on the unconvincing front line that is her living room – 'where I am in Holloway, the violence is coming closer.' 'Shell-shocked', she advises her readers to follow the #riotcleanup hashtag on Twitter.[7] The divide between the university students and the school students, even if now only faintly drawn, could be a sign of worse to come. The video shot during the student protests in which university students grab hold of a boy who'd thrown something burning, and attempt to hand him over to the police, a video which was posted online and subsequently on the Metropolitan Police Wanted list, was reminiscent of those much deeper divides in Paris between the university students and those from the suburbs during the anti-CPE protests. Although NCAFC's statement on the August rioters is more of a patronising ticking off than condemnation, it shows that the student 'spokespeople' fear association with a more uncontainable - unkettlable - battle. It shows the fear, perhaps, that they might be about to lose control.

But it is precisely these more complicated, uncontrollable battles that provide the possibility for something more than a nostalgic reconstruction of the good old days of 2010. As people are evicted from their homes, lose their jobs, are beaten up by police; as their schools are privatised, universities go bankrupt and libraries are destroyed, resistance will have to happen there and then - on a thousand front lines, in street battles and marches, in ongoing strikes and occupations. We will only hear the 'wider echo' of the student protests when anger can no longer be muffled in days out at Westminster, articulated in simple slogans, or separated into causes - when a riot can no longer be just a student riot, and a struggle can no longer be just an education struggle.

FOOTNOTES

1   Educational Maintenance Allowance is a weekly means tested payment of up to £30 given to young people in further education. The scheme was closed to new applicants from England in January 2011.
2   NCAFC website, 'NUS officially supports November 9th national demonstration', 22 September 2011, http://anticuts.com/2011/09/22/nus-officially-supports-november-9th-national-demonstration/ and NUS website, http://www.nus.org.uk
3   NCAFC website, 'NCAFC statement on the riots', 11 August 2011, http://anticuts.com/?s=riots
4   Quoted in John Morgan, 'Pension action plans threaten NUS-UCU alliance', 13 October 2011, http://www.timeshighereducation.co.uk/story.aspsectioncode=26&storycode=417770&c=1
5   Quoted in Toby Helm, 'Unite leader Len McCluskey calls for protests and strikes against cuts', 10 September 2011, http://www.guardian.co.uk/politics/2011/sep/10/unite-len-mccluskey-tuc-strikes
6   NCAFC website, 'NCAFC statement on the riots', 11 August 2011, http://anticuts.com/?s=riots
7   Laurie Penny, 'Panic on the streets of London', 9 August 2011, http://pennyred.blogspot.com/2011/08/panic-on-streets-of-london.html

Sarah Taylor <sarah.taylor55@yahoo.com> has just stopped being a student. She now has nothing better to do than write about student protests

The choice facin

SOCIAL
war
& CLASS CLEANSING
with

Labour
... or back
to hung
council

1 ~~Cle~~
~~you~~

2 Mo

DEMO
3 ~~Imp~~
chil
YUP
4 ~~Doc~~

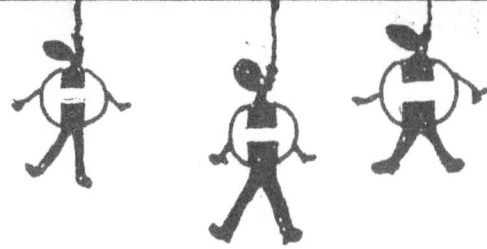

THAMES GATEWAY GHETTOES; BOROUGH REPOPULATE

EVACU

Anonymous, Hackney Council election poster*, May 2006. Note that market-'social' housing 'rent convergence' was already
established Labour policy. 'A cleaner greener Hackney = no more dirty poor people' was reused, e.g. when serving a giant ASBO
on Jules Pipe and Ken Livingstone at the opening of the regenerated Gillett Square as 'Dalston's Covent Garden'.
*The hanged fat bellies are the Hackney Council symbol

Hackney:

ner, greener streets in ~~area~~

NO MORE DIRTY POOR PEOPLE

e police on your streets

HARRY STANLEY R·I·P

~~-ISHED~~ ~~oved~~ schools for your

dren *(HACKNEY WELCOMES THE EDUCATIONAL EXPERTISE OF SWISS BANK U.B.S!)*

E PALACES WHERE

~~nt homes on~~ your estates

USED TO BE —

OFFICIAL POLICY:
ALL 'SOCIAL HOUSING TENANTS TO PAY MARKET RENTS BY 2011

Labour

TION OF WORKING-CLASS HACKNEY TO WITH WEALTHY 'CREATIVE' PROFESSIONALS

UNLIMITED LIABILITY OR NOTHING TO LOSE?

A structurally adjusted version of an article first written for Wildcat magazine in April 2011, in which CLINICAL WASTEMAN *considers crisis management in the world's most 'financialised' and therefore most state-permeated economy – the UK*

April 2011. The UK government, responding to business concern at its plan to impose a fixed limit on non-European immigration, quietly announces some concessions. Permanent residence will be open to anyone bringing £5 million into the country *or able to borrow the same amount from a British bank, secured on assets held elsewhere.* A trivial adjustment of migration policy, but an eloquent statement of the idea of 'national prosperity' underlying 'post'-crisis economic management. The priority is 'controlled' *reflation* of the pre-crisis economy of private credit circulation, appreciating asset prices and associated services along a spectrum from financial to menial.

But no asset boom can be reflated without backing from wholesale asset handlers (misleadingly collectivised as 'markets'), an affinity group whose internationalism (for purposes of arbitrage) and political intransigence the left might learn from. The demands are familiar because they never change: protection of creditors, 'flexiblity' of capital and labour markets, and transfer of the resulting risk (or liability) to the non-asset-owning class. In practice this calls for a show of intent to cut down *without harm to asset prices* the fiscal deficit and public debt swollen by the state's assumption since 2007 of an unpayable £250bn+ debt overhead contracted by... private asset wholesalers.

The 'savings' are supposed to be made by scaling back the state's role as employer of last resort and welfare provider to the large part of the population which the asset boom economy was never able profitably to exploit. But, cutting cash payments to the working class is by no means the same thing as 'shrinking the state'. Any government expecting to maintain order while removing millions of people's legal means of social reproduction must prepare at the same time for *expansion* of the core state function: policing in the broadest sense. One way of stating the stakes of the current and imminent struggles would be to ask: what kinds of 'disorder' (individual-'criminal', 'racial'-sectarian, class-based?) will the state be forced to police, and how successful will the policing be?

The expression 'privatisation of debt', used by some leftists to describe the general policy trend in the UK and elsewhere, is accurate if understood in a double sense. On one level it refers to the transfer of the pre-crisis financial debt overhead from the private sector to the state and onward to the working and assetless class; on another, to the advertised purpose of the effort, the idea that lifting the public debt burden (or rather dumping it downwards) will somehow clear the way for renewed growth of a private sector centred on the creation, expansion and monetisation of credit, i.e. debt.

The inability of this method to restore long term accumulation is obvious enough that various Marxists, shock therapy monetarists, conservative Keynesians and neo-Keynesian radicals are all able to prove it to their own satisfaction. But questions of long term accumulation have little to do with the time scale on which 'policymakers' (consultants, senior civil servants and government) and businesses operate.[1] On the 'pragmatic' level of social management and shareholder value, administrators find little alternative to short term reflation of the credit/service economy, given the (mutually hostile) disruptive capacities of one class with everything staked on the asset boom and another kept dependent on its crumbs. Such is the existing concentration

While the reflationary side of debt privatisation makes headlines, its punitive side stretches right down to the pettiest micro interventions

of capital in the FIRE (finance, insurance, real estate) sector and ancillary services that the kind of manufacturing oriented 'rebalancing' demanded by unions and occasionally promised by ministers would wipe out much of the 'wealth' circulating through service and consumer markets, provoking international capital flight and alienating the 'aspirational' or 'hardworking' demographic knitted tightly into FIRE sector dependency through home and small business 'ownership', private pensions and personal financial investment, compromising years of effort to divide this group from the 'feckless' proletariat singled out for attack under the present policy.[2]

The *political* basis of credit reflation is one reason for emphasising it; another is that the perspective of debt privatisation connects the stakes across the imaginary line between 'public' and 'private' sectors. This matters because official reduction of the scope of conflict to the 'public sector' - an agenda accepted by unions and many 'anti-cuts' campaigners - is an aggressive tool of class decomposition, mobilising 'aspirational' worker-consumer opinion *against* the supposed 'privileges' of state employees and welfare claimants. The inadequacy of 'defending the public sector' as a form of social counterpower will be ever more obvious over the coming months: the 500,000 state workers slated for redundancy will be 'inside' the private sector from the moment they encounter private welfare contractors, as well as in any future employment, which may anyway be the same work they used to do, sold back to outsourcers on freelance or agency terms.

The reflationary side of debt privatisation is visible on a headline story, macro policy level, while the punitive side - the downward transfer of liability - stretches all the way down to the pettiest micro interventions. (The present government follows its predecessor in an obsession with Behavioural Economics, and has set up a dedicated 'Behavioural Insight Team' to propagate 'social norms', i.e. psychological reflexes of individual liability for social problems.) On the macro level the reflationary and punitive aspects intersect almost everywhere, most obviously in the overall scale of fiscal spending cuts (£83 billion), their concentration in working class 'entitlements' (welfare, education, and municipally administered services such as care for children/elderly/disabled), and the avowed reliance on *monetary stimulus* (a former oxymoron which officials have learned to utter earnestly) to offset the ensuing contraction of demand.[3] Other important elements include:

– Transfer of additional tax burden onto low-end consumption through a VAT increase to 20 percent, as part of a general adjustment of the tax system in favour of national competitive advantage in international arbitrage. (As with the public spending cuts, the effect is supposed to be compensated for by low interest rates: i.e. offset for *homeowners* in particular.)

– Automatic enrolment (with a small print 'opt out' provision) of private sector workers in NEST, a stock market pension scheme.

– Further replacement of direct state handling of those facilities regarded as politically unfeasible to shut down altogether (e.g. welfare, garbage collection, medicine) with state funded 'commissioning' of outsourced contractors, i.e. transfer of captive markets to leverage financed ventures. The most ambitious move in this direction may be the reconstitution of the National Health Service as a wholesale buyer of services from 'any willing provider'. Among the first 'providers' to come forward was accountancy and consultancy

multinational, KPMG, indicating both the scale of the prizes available and the scope for sub-sub-contracting.

– A housing reform package openly relished as the urban equivalent of 'Highland Clearances' by some plain speaking Tories. Drastic reduction of housing benefit (rent subsidy paid through the welfare system) and the increase of 'social' (outsourced ex-public) sector rents to 80 percent of market level must be seen as a renewal of a consistent 30-year policy drive to transfer working class income into the private real estate market, i.e. one of the main long term factors in the pre-crisis FIRE bubble and the crisis itself. Political determination to revive this process is also evident in the relaunch of 'Enterprise Zones': the system of subsidised and unregulated urban clearance and redevelopment with which the Thatcher administration began the 30-year cycle.[4]

Other aspects of social punishment, overlapping with those already mentioned in some cases and less widely reported in others, contribute less obviously to reflation or even to fiscal savings; these can be understood as extracting 'payment for the crisis' in a disciplinary rather than monetary sense, or more practically as cultivating the mix of personal desperation and aspiration that a credit-service economy requires of its workers, i.e. the 'basic skills' or 'life skills' whose absence business lobbyists perpetually lament. Examples include:

– Massive transfer of sickness benefit claimants onto the dole and the aggressive 'workfare' programmes attached to it, substantially increasing the number of forced competitors for what all institutional forecasts agree will be a static or falling number of jobs.

– Similarly intensified competition between individual workers outside the welfare system: e.g. public and private sector employers forcing all employees to reapply simultaneously and competitively for reduced numbers of jobs on downgraded terms. Among others, 170,000 municipal workers across the country face an immediate ultimatum to sign new contracts or be fired. (Plus a government promise perfectly encapsulating Behavioural Economists' ideal of 'fairness': all young people will have the *opportunity* to work as unpaid interns.)

– Abolition of legal aid (means tested state contribution to legal fees) for employment, welfare, housing, immigration and clinical negligence cases, i.e. exactly the kind of disputes likely to proliferate in the near future.

– Introduction of fees and access restrictions for employment tribunals (legally binding hearings on unfair dismissal, discrimination etc.); repeal of a large amount of workplace safety law, with funding cut by 35 percent for the body enforcing the remainder; exemption for small businesses from labour legislation. These latter measures reward years of business lobbying and are accompanied by an 'Employer's Charter' 'reminding' bosses of their 'right to ask workers to take a pay cut'.

The same deployment of punitive logic at administrative level is also visible in programmes of marginal scale and ideas yet to be fully implemented:

– A food voucher scheme (run by a Christian charity) for recalcitrant dole claimants whose money is cut off, making it easier for 'welfare-to-work' contractors to stop the payments and hastening the convergence between the general welfare system and the openly 'deterrent' voucher based mechanism for asylum seeking migrants.

– A gap in the all round system of state support for real estate accumulation closed at last by legislation to criminalise squatting, disingenuously helped along by *The Evening Standard*'s stories about *immigrants* trying to squat *already inhabited* buildings. Displacing an existing occupier is of course *already illegal*, but the image of Baltic squatters exploiting 'soft touch' British law in their own sort of international arbitrage gave the panic a momentum of its own.

– A government commissioned Deloitte report recommends compulsory online transactions in all personal interaction with the state (benefit claims, document applications, fee payments), on the grounds that *contributing to cutting 'back-office costs'* (i.e. wages of letter openers and call centre workers) *is a universal social duty*. An all online system would also help to subsidise financial reflation, in that replacement of the former clerical labour would be contracted to the overlapping financial/IT consultancy/services sector (e.g. Deloitte).

– A major speech by David Cameron modifying, though by no means softening, the use of anti-immigration sentiment for purposes of class decomposition. The welfare system, he declared, is 'to blame for creating a generation of work shy Britons, *allowing* migrants to take jobs.' Thus the pretence that immigration control is about protecting native labour from foreign competition is dropped altogether, at the same time as the *social* undesirability of foreign workers is raised to the level of a self-evident premise. Foreign proletarians are *intrinsically* a problem, and 'lazy' British members of the same class are *to blame for it*: therefore the burden of punishment must fall on both. The stakes of this shift in emphasis are high: will it deepen

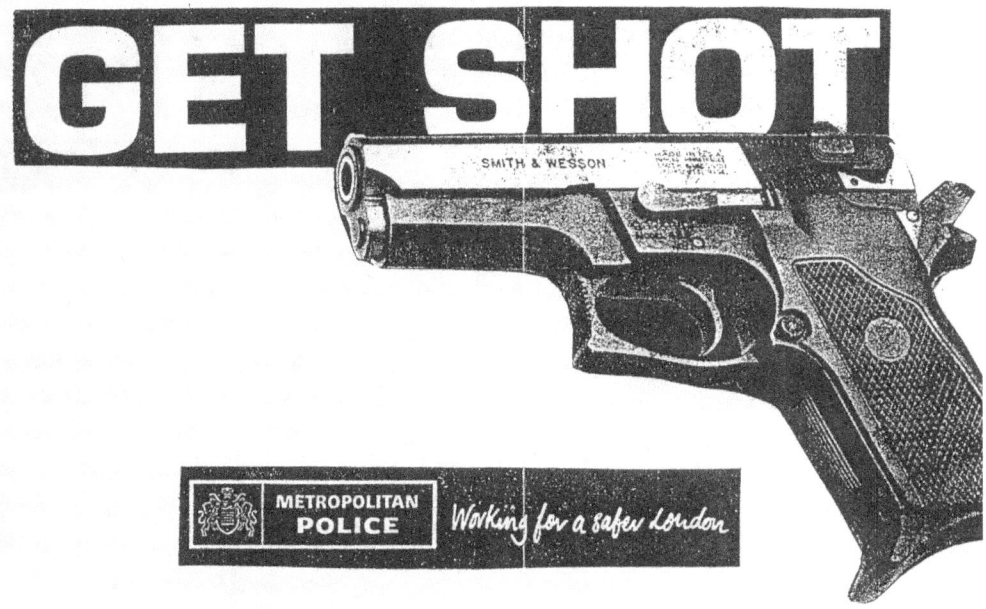

Anonymous, 'Creative Hackney, a Great Place to Live-Work and GET SHOT'. The 'Get Shot' phrase comes from a police gun amnesty campaign, the rest is generic for Hackney c.2004-6. Real estate spokesthings were complaining that 'fear of crime' was the one thing keeping the young and aspirational out (or less overwhelmingly than they already were). Much of the resulting attempt to play on this fear was scrawled directly on walls: 'Luxury Apartments for Crime Victims Opening Soon' etc., plus simple & accurate predictions of the near future: 'Debtors Prison Opening Soon' on high-end developments.

Foreign proletarians are intrinsically a problem, and 'lazy' British members of the same class are to blame for it

hostility between 'British' and 'foreign' workers as the former blame the latter not only for 'stealing jobs', but also for the punitive attacks of state and capital, or might the promise of punishment for all actually contribute to elementary, self-interested solidarity across different 'nationalities' in the same material position?

In April, this account was attached to a description of the social counterpowers then trying to spoil the management model. Because the social work of spoiling is ongoing and volatile, a report on where it stood six months ago is pointless now. But the policing plan has run unchanged through months of bathetic protest gestures and untranslatable riots. In October, the uninventably named Lord Chief Justice Lord Judge threw out the first appeals against post-riot prison sentences, and in so doing explicitly invoked the *collective* nature of the offences as the reason for spectacular punishment:

> The reality is that the offenders were deriving support and encouragement from being together with other offenders and offering comfort, support and encouragement to the other offenders around them. Perhaps too the sheer numbers involved may have led some of them to believe that they were untouchable.

The Clinical Wasteman is certified Hard To Reach. A true address is just somewhere snipers count heads

FOOTNOTES

1   Keynes' 'in the long run we are all dead' has been
quoted often since the crisis broke out, with little
overt acknowledgement that the lifespan implied in
business and political strategy has contracted since
the mid-20th century from human to something
more like feline.

2   (From a forthcoming dictionary):
Aspiration | aspirational. The invention of the
highly elastic adjective 'aspirational' coincides
with what until recently looked like a permanent
shift in the scope of the noun 'aspiration'. Samuel
P. Huntington, in his 1973 rant to the Trilateral
Commission, deplored 'aspiration' as an extravagant
and dangerous *collectively staked* claim: an
overeducated underclass demanding too much and
expecting to get it by forcing *structural* change. But
by the time of its emergence in the 1990s as a
party-political marketing theme, 'aspiration' implied
a strictly *personal* kind of anxious conformism.
The 'aspirational' individual stakes everything on
'social mobility'; that is, she *expects* to compete
against the rest of her class on a 'level playing
field' (i.e. everyone doing the same thing), and she
expects to 'win', beating her opponents 'fairly' by
embracing more eagerly, energetically and obediently
*whatever 'rules of the game' are transmitted from
above*. Or, better still, by correctly *guessing* in
advance the instructions *likely* to be dictated by
previous 'winners' occupying higher rungs on an
imaginary 'career ladder' (comprised in turn of
'playing fields' at ever higher altitudes). This form
of pre-emptive obedience is known as 'showing
initiative'. The curious elasticity of 'aspirational' as an
adjective lies in its multiple applications to a prize
(an 'aspirational' apartment, home entertainment
system, lifestyle), the competitors pursuing it (see
above), and the wider social structure imposing the
competition (an 'aspirational society'). This last usage
confirms the irreconcilable contradiction between
'aspirational' conditions and the kind of aggressive
class 'aspiration' feared by Huntington. Diligent
'aspirational' behaviour precludes the very thought of
provoking structural change, as the existing structure
is the context, vehicle and measure of personal
'success'; turning the world upside down would make
a mockery of the effort to 'rise to the top'.

3   In the sub-zero-sum game of traditional capitalist
crisis management, 'stimulus' always referred to
fiscal policy: the state spending money raised
through borrowing or sometimes even taxation
in ways thought to 'stimulate' (hence the name)
production or just consumer demand. Monetary
policy – central bank manipulation of the number
of currency units chasing each commodity – was
preferred by *opponents* of stimulus. The hybrid
monster 'monetary stimulus' slinks into view
when fiscal spending is political poison but
'Something must, nonetheless, be Seen To Be Done'.
In practice, this means 'grassroots' borrowing
becomes compulsory as newly conjured euros/
dollars/pounds flood the credit system and erode
the value of cash. If all goes well (in a world of
international competition) the inflation – along with
the 'stimulus' – flows with the fresh hot money
out of the stimulant states and into high yielding
'emerging markets'.

4   On the new Enterprise Zones plan in relation to
the old see, Stephen Alexander, 'Enterprise zones
introduced across England', http://www.wsws.org/
articles/2011/apr2011/zone-a26.shtml. On the
old ones in relation to State-led 'leveraging' of
real estate opportunities, see Anna Minton, *Ground
Control*, London: Penguin, 2009.

LOGISTICS AND OPPOSITION

'Sabotage the social machine'. 'Incinerate the documents!' In the first in this issue's series of articles linking logistics, workplace surveillance and national security, ALBERTO TOSCANO examines the anti-urbanist presuppositions of insurrectionary anarchism. Instead of breaking the lines of circulation, he writes, shouldn't radicals imagine repurposing them to entirely new ends?

THE SPONTANEOUS PHILOSOPHY OF INTERRUPTION

It is rare, in contemporary oppositional thought, to encounter the totalising temporal imaginary of revolution that so marked the visions and strategies of the modern left. When it hasn't been victim to melancholy retreats from the teleology of emancipation, that encompassing horizon of social change and political action has come under attack, alongside the very idea of transition, for domesticating antagonism. Interstitial enclaves or temporary liberated zones, ornamented by discourses of withdrawal and difference, have widely replaced the reference to an advancing, unifying and largely homogeneous planetary movement of liberation. The space-time of much of today's anti-capitalism is one of subtraction and interruption, not attack and expansion.

Needless to say, any negation of the status quo brings with it spatial separation and temporal disruption, but the contemporary ideology, or spontaneous philosophy, of interruption appears – perhaps as a testament to the claustrophobia of our present – to make something of a fetish out of rupture. This cuts across theory and activism, laying bare a shared structure of feeling between the political metaphysics of events or 'dissensus' and the everyday tactics of struggles. Foregrounding interruption implies a particular understanding of the nature of contemporary capital, the capabilities of antagonism and the temporality (or lack thereof) of transformation.

The Coming Insurrection formulates, in a compellingly abrasive way, a widespread conviction that contemporary struggles against capital have shifted from the point of production to those of circulation, distribution, transport and consumption. In other words, that arresting the flow of this homogenised society is a *conditio sine qua non* for the irruption of non-capitalist forms-of-life:

> The technical infrastructure of the metropolis is vulnerable. Its flows amount to more than the transportation of people and commodities. Information and energy circulate via wire networks, fibres and channels, and these can be attacked. Nowadays sabotaging the social machine with any real effect involves reappropriating and reinventing the ways of interrupting its networks. How can a TGV line or an electrical network be rendered useless?[1]

Behind this statement lies an anti-urbanism that regards contemporary spectacular exploitation and conformity as products of the capillary management of everyday life. Cities are stripped of any life not mobilised for the commodity and pre-empted from any behaviour at odds with a tautological drive for systemic reproduction:

> The metropolis is not just this urban pile-up, this final collision between city and country. It is also a flow of being and things, a current that runs through fiber-optic networks, through high-speed train lines, satellites, and video surveillance cameras, making sure that this world keeps

like Alice's red queen, by great exertion and utmost speed the metropolis barely manages to remain in the same position

running straight to its ruin. It is a current that would like to drag everything along in its hopeless mobility, to mobilize each and every one of us.[2]

The interruption or sabotage of the infrastructure of mobilisation are the other side of *The Coming Insurrection*'s conception of communes not as enclaves for beautiful souls, but as experiences through which to develop the collective organs to both foster and endure the crisis of present, and to do so in a fashion that does not sever means from ends. The book's catastrophic optimism lies in advocating that interruption is somehow generative of anti-capitalist collectivity (rather than passing irritation or mass reaction). It is also founded on a repudiation of the inauthenticity of massively mediated, separated and atomised lives in the metropolis.

There are inadvertent echoes of Jane Jacobs in the scorn against 'indifferent' modern housing and the idea that with 'the proliferation of means of movement and communication, and with the lure of always being elsewhere, we are continuously torn from the *here and now*'.[3] Real communities that do not rest on the atrophying of bodies into legal identities and commodified habits are to emerge out of the sabotaging of *all* the dominant forms of social reproduction, in particular the ones that administer the ubiquitous mobilisation of 'human resources'. Materialism and strategy are obviated by an anti-programmatic assertion of the ethical, which appears to repudiate the pressing critical and realist question of how the structures and flows that separate us from our capacities for collective action could be turned to different ends, rather than merely brought to a halt.

The spatial vocabulary articulated in *The Coming Insurrection* is, to employ a well worn dichotomy, not one of revolution but one of revolt. This spatial distinction between negations of the status quo was beautifully traced through the relationship between Rimbaud and the Paris Commune by the Italian critic Furio Jesi. Jesi begins with the evident temporal distinction between revolution conceived in terms of the conscious concatenation of long- and short-term actions aimed at systemic transformation in historical time and revolt as a suspension of historical time. Revolt is not the building up but the revelation of a collectivity. It is, to borrow from André Malraux's *Hope*, an organised apocalypse.

In this abrogation of the ordered rhythms of individual life, with its incessant sequence of personal battles, revolt generates 'a shelter from historical time in which an entire collectivity finds refuge'.[4] But the interruption of historical time is also the circumscription of a certain a- or anti-historical space, a space torn from its functional coordinates:

Until a moment before the clash [...] the potential rebel lives in his house or his refuge, often with his relatives; and as much as that residence and that environment may be provisional, precarious, conditioned by the imminent revolt, until the revolt begins they are the site of an individual battle, more or less solitary. [...] You can love a city, you can recognise its houses and its streets in your most remote and secret memories; but only in the hour of revolt is the city really felt like an haut-lieu [a high place] and at the same time your own city: your own because it belongs to you but at the same time also to others; your own because it is a battlefield you and the collectivity have chosen; your own, because it is a circumscribed space in which historical time is suspended and in which every act has its own value, in its immediate consequences.[5]

The collective experience of time, and of what Jesi calls symbols (such that the present adversary simply becomes *the enemy*, the club in my hand *the weapon*, victory *the just act*, and so on), means that the revolt is an action for action's sake, an end (as in The Invisible Committee's reflections on the ethics of sabotage and the commune) inseparable from its means.

It was a matter of acting once and for all, and the fruit of the action was contained in the action itself. Every decisive choice, every irrevocable action, meant being in accordance with time; every hesitation, to be out of time. When everything came to an end, some of the true protagonists had left the stage forever.[6]

Abiding with the interruptive paradigm of an intransitive and intransigent revolt, we can wonder whether, and if so to what extent, the historical space that revolt intervenes in inflects its character. It is no accident that the kind of sabotage envisioned in *The Coming Insurrection* is on lines and nodes of circulation, and not on the machinery of production itself.

THE TRIUMPH OF PROCESSING

The centrality to an intensely urbanised capital of the efficient, profitable, ceaseless and standardised movement of material and information – the very target of *The Coming Insurrection*'s ethics of interruption – has been noted for a long time. Fifty years ago, Lewis Mumford, writing in *The City in History* of the catastrophic propensities of the contemporary metropolis – what he elegantly called 'the aimless giantism of the whole' – pointed to the pivotal role of the growing possibilities of supply to the 'insensate agglomeration of populations' in exponentially expanding cities, and their

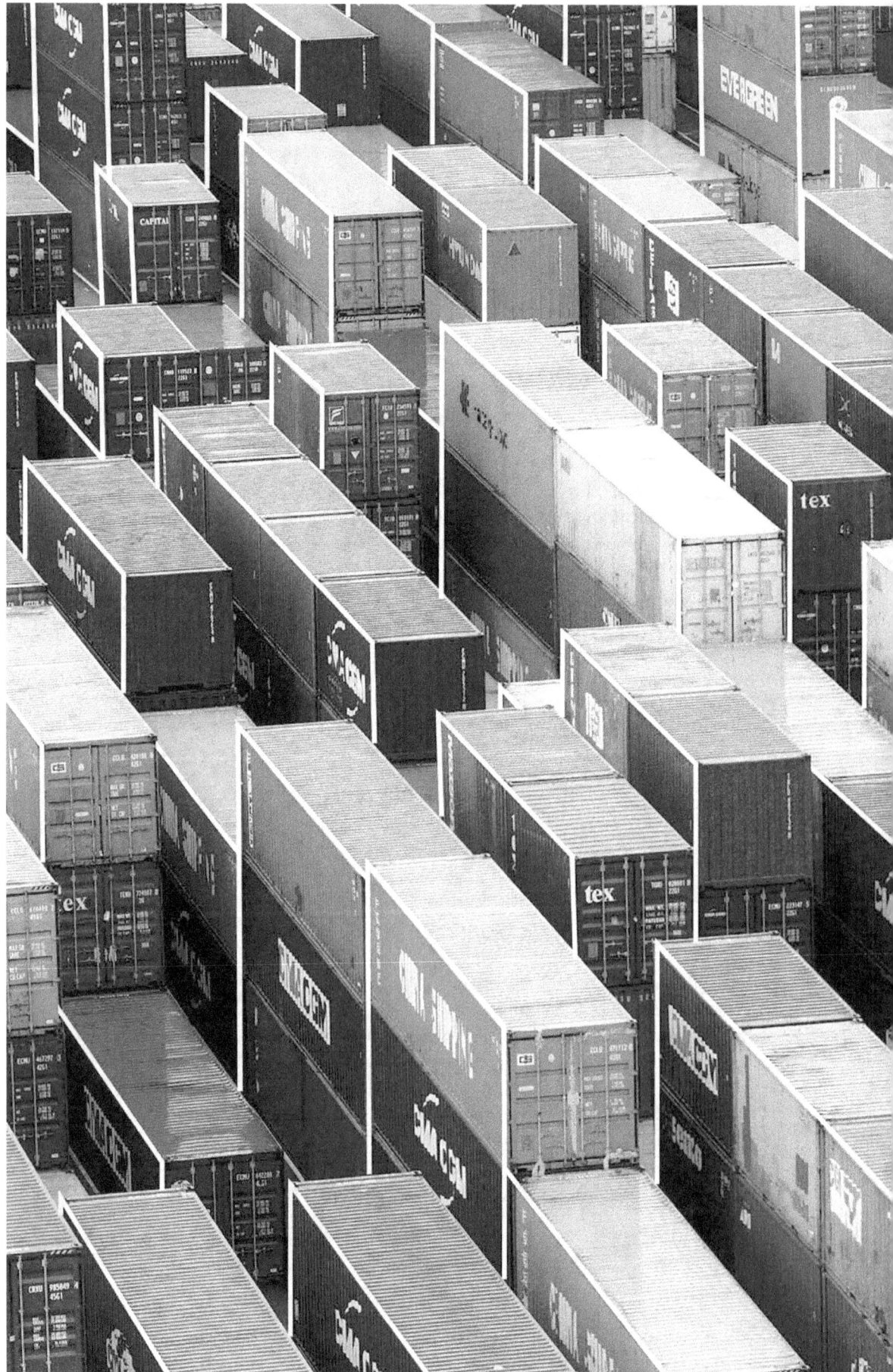

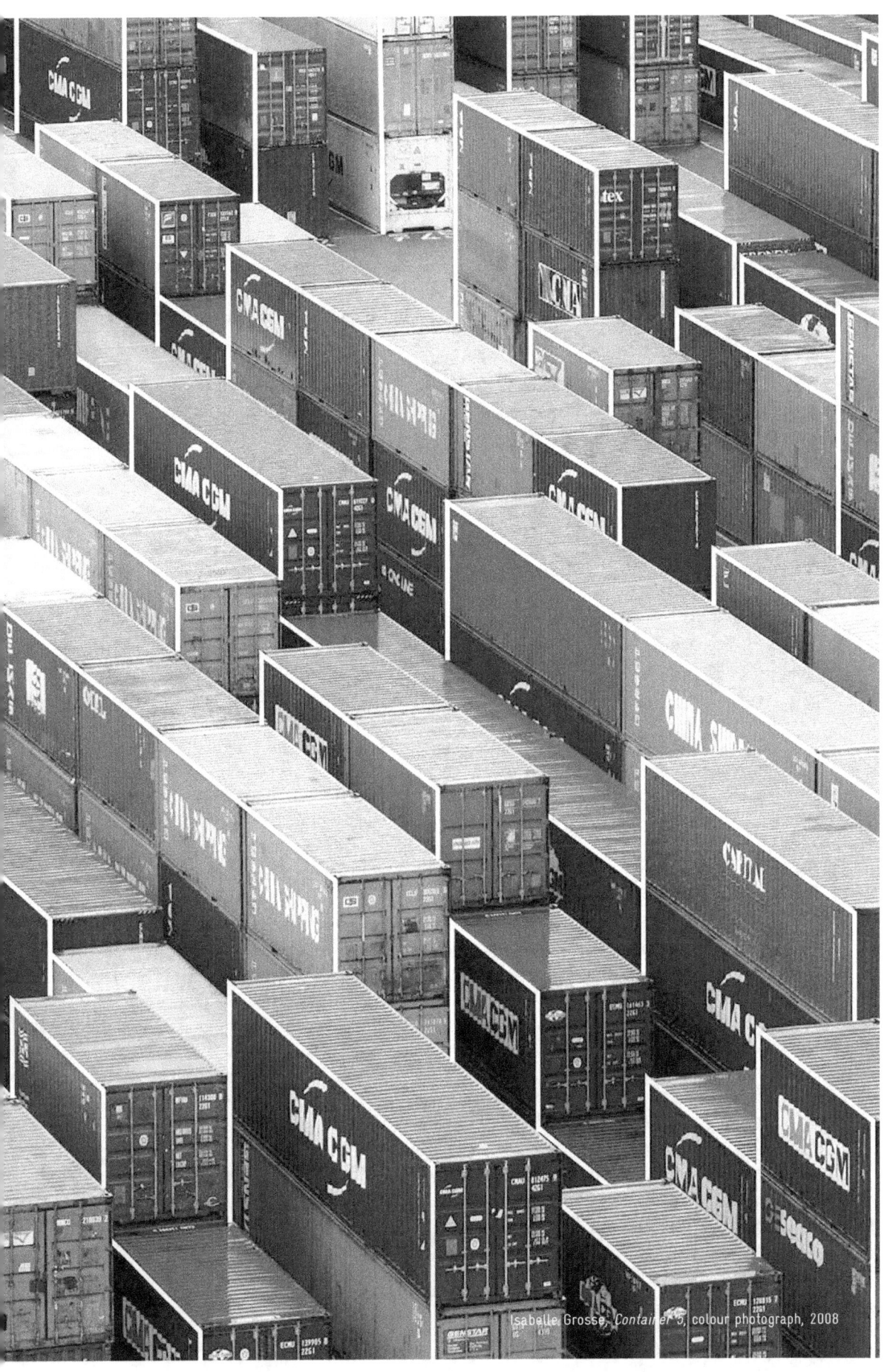

Isabelle Grosse, *Container 5*, colour photograph, 2008

relations to the 'tentacular bureaucracies' that controlled such flows of goods.

> During the 19th century, as populations heaped further into a few great centres, they were forced to rely more fully on distant sources of supply: to widen the basis of supplies and to protect the 'life-line' that connects the source with the voracious mouth of the metropolis, became the function of army and navy. In so far as the metropolis, by fair means or foul, is able to control distant sources of food and raw materials, the growth of the capital can proceed indefinitely.[7]

The organisational and energetic resources required to reproduce the metropolis are formidable: 'like Alice's red queen, by great exertion and utmost speed the metropolis barely manages to remain in the same position.'[8] The metropolis has the intensification and expansion of supply lines as its precondition, and logistics becomes its primary concern, its foremost product, and the basic determinant of its power:

> The metropolis is in fact a processing centre, in which a vast variety of goods, material and spiritual, is mechanically sorted and reduced to a limited number of standardized articles, uniformly packaged, and distributed through controlled channels to their destination, bearing the approved metropolitan label. 'Processing' has now become the chief form of metropolitan control.[9]

Despite his systemic objections to the catastrophic ends of this amorphous machine for (capital) accumulation, Mumford also regards these control capabilities as potentially reconfigurable in a multi-centred and organic society. But, especially when it comes to the informational requirements attendant on such control-by-processing, manifest in the metastasis of a tentacular bureaucracy, he too is tempted by the possibilities of insurgent interruption – even recalling an anarchist slogan ('*Incinerate the documents!*') to stress the ease with which such a system, founded on the circulation of real or virtual 'paper', could be ground to a halt.

But it is also possible, and indeed necessary, to think of logistics not just as the site of interruption, but as the stake of enduring and articulated struggles. Here there remains much to digest and learn from in the ongoing research of labour theorist and historian Sergio Bologna, an editor in the 1970s of the journal *Primo Maggio*, which carried out seminal inquiries into containerisation and the struggles of port workers.[10] Countering those 'post-workerists' who have equated post-Fordism with the rise of the cognitive and the immaterial (and basically with the ubiquity of a figure of work patently traced on that of the academic or 'culture worker'), Bologna notes that the key networks that condition contemporary capitalism are neither affective nor simply digital, but involve instead the massive expansion and constant innovation in the very material domain of logistics – in particular of 'supply chain management', conceived of in terms of the speed, flexibility, control, capillary character and global coverage of the stocking, transport and circulation of services and commodities.[11]

Bologna underscores the military origins of logistics, namely in the work of de Jomine, a Swiss military theoretician working first under Napoleon and then under the Russian Tsar Alexander I. The 'original function of logistics', writes Bologna,

> was to organise the supplying of troops in movement through a hostile territory. Logistics is

not sedentary, since it is the art of optimizing flows [...] So logistics must not only be able to know how to make food, medicines, weapons, materials, fuel and correspondence reach an army in movement, but it must also know where to stock them, in what quantities, where to distribute the storage sites, how to evacuate them when needed; it must know how to transport all of this stuff and in what quantity so that it is sufficient to satisfy the requirements but not so much as to weigh down the movement of troops, and it must know how to do this for land, sea and air forces.[12]

He goes on to analyse how the problems of logistics have been central to the ongoing transformations of contemporary capitalism, from the just-in-time organisation of production of 'Toyotism', to the world-transforming effects of containerisation (itself accelerated by its military-logistical use in the Vietnam War).[13] The homogenisation registered at an existential level by *The Coming Insurrection* is here given a very prosaic but momentous form in the standardisation and modularisation that characterises a planetary logistics which, in order to maintain the smoothness and flexibility of flows, must abstract out any differences that would lead to excessive friction and inertia.

For my purposes, however, what is paramount is what this logistical view of post-Fordism tells us about the character of antagonism, and specifically of class struggle. Narcissistically mesmerised by hackers, interns and precarious academics, radical theorists of post-Fordism have ignored what Bologna calls 'the multitude of globalisation', that is all of those who work across the supply chain, in the manual and intellectual labour that makes highly complex integrated transnational systems of warehousing, transport and control possible. In this 'second geography' of

It is possible, and indeed necessary, to think of logistics not just as the site of interruption, but as the stake of enduring struggles

logistical spaces, we also encounter the greatest 'criticality' of the system – though not, as in the proclamations of *The Coming Insurrection*, in the isolated and ephemeral act of sabotage, but in a working class which retains the residual power of interrupting the productive cycle – a power that offshoring, outsourcing, and downsizing has in many respects stripped from the majority of 'productive' workers themselves.

Here it is possible to link the question of logistics quite closely to that of the management of labour and the neutralisation of class struggle, in a way that sheds some doubt on the 'criticality' identified by Bologna. The expulsion of a mass labour force from containerised ports, their physical separation from zones of urbanisation and connection to other labourers, as well as the deeply divisive labour regulations that divide an international maritime labour force are an important instance of this. As Tim Mitchell writes in his fine essay on energy and the spatial history of class struggle, 'Carbon Democracy':

> Compared to carrying coal by rail, moving oil by sea eliminated the labour of coal-heavers and stokers, and thus the power of organized workers to withdraw their labour from a critical point in the energy system [...] [W]hereas the movement of coal tended to follow dendritic networks, with branches at each end but a single main channel, creating potential choke points at several junctures, oil flowed along networks that often had the properties of a grid, like an electrical grid, where there is more than one possible path and the flow of energy can switch to avoid blockages or overcome breakdowns.[14]

REFUNCTIONING THE SPACES OF CAPITAL

The electrical grid provides an apt transition to reflecting on the relationship between the logistics of capital and the spatial politics of anti-capitalism in a manner that does not merely involve the bare negation or mere sabotage of the former by the latter. The power grid (contrasted with the railway network) was in fact a system whose capabilities for coordinated decentralisation were emphasised by Mumford as a necessary model for a shift out of an aimlessly urbanising capitalism. Following Mumford, a number of Marxist theorists have of late reflected – in a mode that, to borrow a recent quip from David Harvey, we could call *pre-communist* rather than post-modern – on what aspects of contemporary capitalism could be refunctioned in the passage to a communist society. Obversely to *The Coming Insurrection*, they have asked how could a high-speed rail system or an electrical network be rendered not useless, but *useful* – in what would clearly need to be a thoroughly redefined conception of use, one not mediated and dominated by the abstract compulsions of value and exchange.

It is striking that many of these authors have put logistical questions at the forefront of these thought experiments, almost as though logistics were capitalism's *pharmakon*, the cause for its pathologies (from the damaging hypertrophy of long-distance transport of commodities to the aimless sprawl of contemporary conurbation) as well as the potential domain of anti-capitalist solutions. In this vein, Fredric Jameson has recently, and somewhat perversely, identified the distribution systems of Wal-Mart, the very emblem of capitalism's seemingly inexhaustible capacity for devastating mediocrity, as precisely one of those aspects of capitalism whose dialectical refunctioning, or whose change of valence, could give a determinate character to our social utopias.[15]

The ambivalence of logistics, and particularly of the environmental consequences

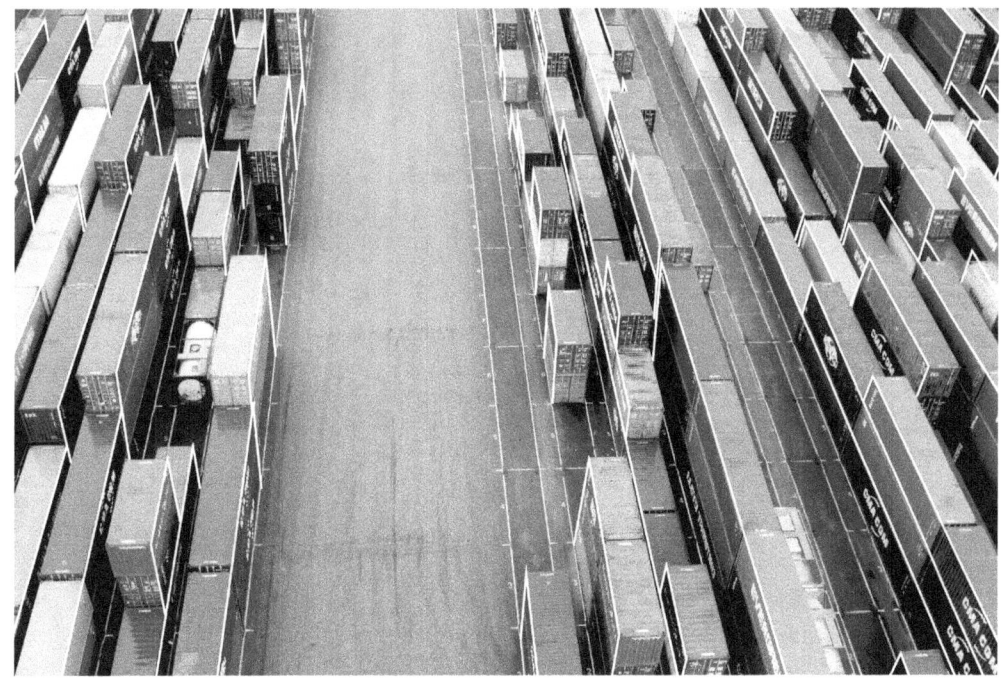

Isabelle Grosse, *Container 2*, colour photograph, 2008

of the unprecedented logistical and energetic complexes that make contemporary megalopolises both the drivers and the possible sites for a response to catastrophic climate change (among other processes) have led Mike Davis, in his appropriately titled 'Who Will Build the Ark?', to demand that, recalling the great experiments in urbanism of the USSR in the 1920s, we begin to look for the potentialities for a non-capitalist and non-catastrophic future in cities themselves.[16] In particular, Davis has advanced, to borrow from Mitchell, some of the parameters of a low-carbon democratic socialism. Arguing, contrary to the Malthusianism of much of the green movement that it is 'the priority given to public affluence over private wealth' that can set the standard for a conversion of engines of doom into resources of hope.

As Davis writes:

Most contemporary cities repress the potential environmental efficiencies inherent in human-settlement density. The ecological genius of the city remains a vast, largely hidden power. But there is no planetary shortage of 'carrying capacity' if we are willing to make democratic public space, rather than modular, private consumption, the engine of sustainable equality.[17]

Such an assertion of the necessity of a drastic transition, as against plural but ineffectual interruptions, takes logistic and energetic dimensions of anti-capitalist struggle more seriously than the convergence of anti-urbanist visions of space and epiphanic models of revolt that – for evident and in many respects sacrosanct historical and political reasons –

have come to dominate much anti-capitalist thought.[18] It also does so by recognising what, by analogy with Herbert Marcuse, we could call the *necessary alienation* involved in complex social systems, including post-capitalist ones. As David Harvey has noted, against the grain of fantasies of a tabula rasa, unmediated communism or anarchism:

> The proper management of constituted environments (and in this I include their long-term socialistic or ecological transformation into something completely different) may therefore require transitional political institutions, hierarchies of power relations, and systems of governance that could well be anathema to both ecologists and socialists alike. This is so because, in a fundamental sense, there is nothing unnatural about New York city and sustaining such an ecosystem even in transition entails an inevitable compromise with the forms of social organization and social relations which produced it.[19]

The question of what use can be drawn from the dead labours which crowd the earth's crust in a world no longer dominated by value proves to be a much more radical question, and a much more determinate negation than that of how to render the metropolis, and thus in the end ourselves, useless.

Alberto Toscano <sos01at@gold.ac.uk> teaches at Goldsmiths, University of London and sits on the editorial board of *Historical Materialism*. He is the author of *Fanaticism: On the Uses of an Idea*, London: Verso, 2010

NOTE

Versions of this article were originally presented at the Spaces of Alterity conference at University of Nottingham, and The Anarchist Turn colloquium at the New School for Social Research. I'm grateful to the organisers for providing me with the occasion to hone some of these ideas. Thanks to Jason Smith, Joshua Clover, Evan Calder Williams, Benjamin Noys, Josephine Berry Slater and Benedict Seymour for references, critique and inspiration.

FOOTNOTES

1 The Invisible Committee, *The Coming Insurrection*, Los Angeles: Semiotext(e), 2009, pp.111–12. These reflections prolong those initially spurred by the so-called Tarnac affair, which saw this anonymous argument for sabotage transformed into the flimsy basis for a prosecutorial campaign at once vicious and spurious. See my 'The War Against Pre-Terrorism', *Radical Philosophy*, issue 154, 2009, pp.2–7, http://www.radicalphilosophy.com/commentary/the-war-against-pre-terrorism .

2 *The Coming Insurrection*, pp.58–9.

3 Ibid., p.59.

4 Furio Jesi, *Lettura del 'Bateau ivre' di Rimbaud* (1972), Macerata: Quodlibet, 1996, p.22.

5 Ibid., pp.23–4.

6 Ibid., p.24.

7 Lewis Mumford, *The City in History*, New York: Harcourt, 1961, p.539.

8 Ibid., p.540.

9 Ibid., pp.541–2.

10 For an excellent introduction to the work of Bologna and *Primo Maggio* in English, which stresses the way in which it both prolonged and challenged *operaismo* through a historiographic lens, see Steve Wright, *Storming Heaven: Class Composition and Struggle in Italian Autonomist Marxism*, London: Pluto, 2002, chapter 8: 'The Historiography of the Mass Worker'. The full collection of *Primo Maggio* is now available as a CD-ROM in *La rivista Primo Maggio* (1973-1989), Cesare Bermani (Ed.), Rome: DeriveApprodi, 2010.

11 For an incisive and informed treatment of the logistics revolution and the challenge it poses to workerist and autonomist perceptions of class struggle, see Brian Ashton, 'The Factory Without Walls', *Mute*, http://www.metamute.org/en/Factory-Without-Walls. Ashton underscores the link

between any future resurgence of oppositional anti-capitalist organising and knowledge of capital's composition and operation – the cognitive mapping of supply chains, value extraction and the levers of struggle.

12 Sergio Bologna, 'L'undicesima tesi', in *Ceti medi senza futuro? Scritti, appunti sul lavoro e altro*, Rome: DeriveApprodi, 2007, p.84.

13 See Marc Levinson, *The Box: How the Shipping Container Made the World Smaller and the World Economy Bigger*, Princeton: Princeton University Press, 2008, chapter 9.

14 Timothy Mitchell, 'Carbon Democracy', *Economy & Society* 38.3, 2009, p.407.

15 Fredric Jameson, 'Utopia as Replication', in *Valences of the Dialectic*, London: Verso, 2009, pp.420–434.

16 An interrogation of the logistical dimensions of transition, state building and class struggle in the USSR would need to take its cue from chapter four of Robert Linhart's arresting study of the conjunctural and contradictory character of Lenin's thought and politics post-1917, *Lénine, les paysans, Taylor*, re-edited by Seuil in 2010 – a book quite unique in its combination of a real appreciation of Lenin with a welcome rejection of the comforting apologias of Leninism. This chapter, entitled 'The railways: the emergence of the Soviet ideology of the labour-process', details how, in the context of the famine, the authoritarian Taylorist turn in the organisation of work was driven through in that sector which provided the vital hinge between production, services and administration, and whose critical disorganisation was exacerbated by the very autonomous workers' organisation that had previously made it into a hub of anti-Tsarist organising, and now appeared as a kind of economic blackmail all the more menacing in that it took place within the crisis of the civil war. The Bolsheviks, he notes, were 'almost instinctively attentive to everything that concerns communication, flow, circuits' (p.151). In this moment, the railways appeared as the nerve fibres and life blood of a 'state in movement', and militarised centralisation, planning and labour discipline as imperatives – as evidenced, among others, by Trotsky's 'order 1042', viewed by Linhart as the first key instance of state planning. After all, 'if there is an activity that must, by nature, function as a *single mechanism*, one that is perfectly regulated, standardised and unified throughout the country, it's the railway system' (p.162). The

seemingly inevitable Taylorisation of the railways both forges and deforms the USSR, especially in furthering the split, thematised by Linhart, between the proletarian as political subject and as object of iron discipline. Among the more interesting sites of the necessary fixation on logistics (namely, on railways and electrification) are the films of Dziga Vertov, which promise a cognitive mapping that would join the Taylorist decomposition of labour, imaged as 'a regular, uninterrupted flow of communication', and its subjective mastery, in which the 'transparency of the productive process' (p.169) is provided to each worker in the guise of an all encompassing vision.

17 Mike Davis, 'Who Will Build the Ark?', *New Left Review* II/61 (2010), p.43.

18 For a transitional proposal or 'determinate negation of the existent' that stakes some of the same ground as Davis and Harvey, albeit from a different Marxian vantage, see Loren Goldner, 'Fictitious Capital and the Transition Out of Capitalism', http://home.earthlink.net/~lrgoldner/program.html . In his inventory of transitional negations and the refunctioning of 'total existing means of production and labour power', now grasped as 'use values', Goldner advocates the 'integration of industrial and agricultural production, and the breakup of megalopolitan concentration of population. This implies the abolition of suburbia and exurbia, and radical transformation of cities. The implications of this for energy consumption are profound'. In a logistical vein, he proposes the 'centralization of everything that must be centralized (e.g. use of world resources) and decentralization [of] everything that can be decentralized (e.g. control of labour process within the general framework)'.

19 David Harvey, *Justice, Nature and the Geography of Difference*, Oxford: Blackwell, 1996, p.186.

WHAT THE RFID IS THAT?

In his contribution to this issue's series of articles linking logistics, workplace surveillance and national security, BRIAN ASHTON zooms in on the microscopic technologies surveilling and shaping working lives detractors and defenders

You already have zero privacy. Get over it.
– Scott McNally, Chief Executive of Sun Microsystems[1]

On 13 November, 1964, the acting chief constable of Liverpool announced the formation of a commando force of 'just over 100 specially selected men and women from the city police force who would mix in disguise with the general public.'[2] They would work in conjunction with television cameras that were 'scanning the streets and transmitting pictures to a monitor in headquarters. Any suspicious behaviour would be passed on by radio to the commandos for quick action.'[3] The two-way radios were the first to be issued to a British police force. And the use of cameras was watched with interest by police forces around the country. Of course they sometimes got it wrong:

Disguises were so successful at first that more than once, one team of commandos spent more than 45 minutes watching another team, whom they thought were acting suspiciously, and only found out their real identity when they tried to move them on.[4]

The powerful have always spied on us. They do it to exploit us, to control us, and because they fear us.

With the advent of the modern industrial group in large factories in urban areas, the whole process of control underwent a fundamental revolution. It was now the owner or manager of a factory [...] The employer as he came to be called, who had to secure, or exact, from his employees a level of obedience and/or co-operation which would enable him to exercise control.[5]

In order to do this, the boss man had to employ large numbers of supervisors and company spies, and the spies didn't confine themselves just to the workplace. During the early days of the Ford Motor Company, the Five Dollar Day period, the company's sociological department investigated the social lives of the workers, using 30 investigators to check that the workers were living moral lives. If it was found that the money was 'more of a menace than a benefit to him', that a worker had 'developed weaknesses', his bonuses would be lost for a period of six months. And, if that didn't bring him to his senses, he was sacked.[6] It was, perhaps, an early if crude attempt to connect mentally and emotionally with the workers – like being called in for your monthly assessment. Ford also employed 3,500 security men, whose jobs involved spying on the workers and, if necessary, physically intimidating them, usually by smashing them over the head with a baton. But in the end it was the Fordist mode of production that got beaten over the head with a working class baton. Some 70-odd years after the founding of the Ford Motor Company, the death knell for Fordist methods was sounded by Gianni Angelli, patriarch of the Fiat organisation, when he described Turin's Mirafiori factory as 'the ungovernable factory'.

The defeat of the Fordist mode of production did not mean the death of scientific management, a system that had its genesis in the work of Andrew Ure and Charles Babbage, and whose most famous practioner was Frederick Winslow Taylor. Scientific management:

is an attempt to apply the methods of science to the increasingly complex problems of the control

EPC RFID chip used by Wall Mart

of labor in rapidly growing capitalist enterprises. It lacks the characteristics of a true science because its assumptions reflect nothing more than the outlook of the capitalist with regards to the conditions of employment [...] It does not attempt to discover and confront the cause of this condition, but accepts it as an inexorable given, a 'natural' condition. It investigates not labor in general, but the adaptation of labor to the needs of capital.[7]

Scientific management takes technology as a given – and in this period it should be taken as a given that scientific management is an integral part of the technology that confronts us in our daily life. This article is an attempt to look at how capital tries to use technology to control us 24/7.

Along the continuous supply chains of 21st century capitalism reside technological apparatuses that have the capacity to gather information on you. You are policed in ways you may not be aware of; your work uniform could be telling tales on you this very minute, and, if you are using your work computer to read this article on the *Mute* website, it may well be filming you doing so. Back to work, pal, now. And if you're tardy in responding to that order, your tardiness could well be recorded by that clock, up there on the wall. Because not only does it go 'tick tock', it also films you and records your conversations. And can do so for up to 21 days.

Spying on workers is big business; the net abounds with companies producing and selling the tools to spy on you and keep you in line. The use of workplace technologies by workers for their own benefit is widespread and forces capital to seek the means to curtail it. One company, Spectorsoft, sells a piece of software called Spector CNE Investigator – an employee investigation system. One of the case studies used to advertise the product explains how an electrical utility company used the system to control the use of its computer networks. It quotes the utility company's information security officer:

> Every company will have a need to investigate inappropriate behaviour from time to time. Spector CNE is the complete tool to obtain the needed information. If we ever have to go to court, CNE's screen snapshots are irrefutable [...] Finding and weeding out a problem employee early, that really saves the company money [...] Now I know EXACTLY what an individual is doing. I get to see it like a videotape.[8]

The company mentioned here is based in California, but such software is available globally. Last year in Liverpool, three workers in a benefits office were sacked for inappropriate use of computers. Imagine such action being

---

The powerful have always spied on us

carried out across the entire Benefits Agency; it could save a lot of money from being paid out in redundancy packages. The companies that produce and sell surveillance equipment also make commodities that can stop such equipment working. Isn't capitalism wonderful?

Spying on workers is widespread and, in some cases, impacts on thousands of people. In 2009 numerous high profile companies in Germany were found to be spying on their workers.[9] An employee at Deutsche Telekom had his mobile phone records checked by the company during an investigation into a leak of information to the media. The man was a union representative. On a much larger scale, the rail company Deutsche Bahn owned up to spying on its workforce. Their activity included the monitoring of emails, checking on how many toilet breaks were being taken by individuals, and prying in to the love lives of workers. In 2002-2003, 173,000 workers were screened; in 2005 the entire workforce of 220,000 was spied on.[10] The head of Deutsche Bahn, Hartmut Mehdorn, was forced to resign over the issue. Perhaps they got the idea from the retail industry, because in 2008 two discount companies, Lidl and Schlecker, were exposed in the mainstream media for widespread spying. The German weekly, Stern, said it had obtained hundreds of Stasi-like logbooks that minutely monitored the movements and private conversations of employees.

What is happening now is that different aspects of technology are being linked up to produce overarching systems of surveillance. Global Positioning Systems (GPS) and Radio Frequency Identification tags (RFID) are being used to spy on workers in and out of the workplace. GPS relies on satellites orbiting hundreds of miles above the earth. A GPS unit is fitted to a vehicle; this unit can correspond with a satellite and enable the tracking of the vehicle and its driver. In the USA for example there have been instances where GPS technology has been used to monitor workers involved in union recruiting campaigns. In one case, the only two company vehicles fitted with the equipment were the ones used by two workers believed to be organising such a campaign. The National Labour Relations Board (NLRB) recognised that GPS technology would allow the company to interfere with its employees' protected concerted activity by tracking the two worker-organisers' every move in real time to immediately detect, for example, if they met at the same location or visited other employees

at their homes during non-work hours.[11] An RFID consists of a microchip and an antenna and is activated by radio waves transmitted by a reader/scanner. Some tags have minute batteries attached and can self-activate.

In the end Fordism got beaten over the head with a working class baton

Although GPS and RFID are technologically different, they share some common abilities. Both can track the locations of individuals, company vehicles and cargos in real time, creating electronic logs relating to location and movement, which can be used to generate detailed reports that could be used to discipline or sack workers. Both systems are out of sight, making them easy to forget – and we do forget. How many people remember the construction of the world's largest road and vehicle surveillance system? The system that was put in place in Britain over a five-year period starting in 1999 resulted in the installation of thousands of number plate recognition cameras. Cameras that can communicate with the Police National Computer in real time through microwave links and the telephone system.[12]

Spy cameras are now ubiquitous; they are an integral part of the modern architectural landscape. But don't worry, they are there for your protection. Honest.

RFID is the technology that has the potential to be the most pervasive. It can be put anywhere, and that includes inside you. How does it work? They inject it in to your arm. This is the technical bit. It is a simple construction that consists of a coil of wire and a microchip, hermetically sealed inside of a glass capsule.

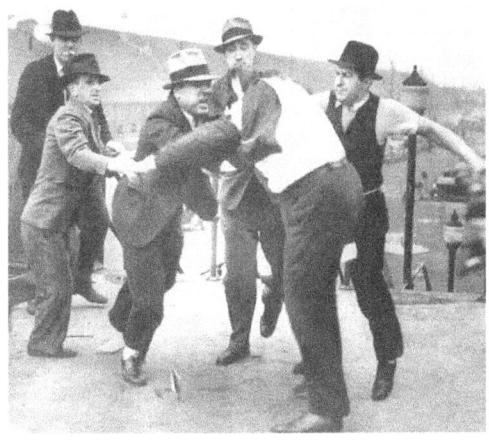

Ford Motor Company security guards attack UAW union organisers, Detroit 1937

These devices are 11 millimetres long and about one millimetre in diameter, comparable to a grain of rice. The coil acts as an antenna and uses an RFID reader/scanner's varying magnetic field to power the microchip and transmit a radio signal. The chip modulates the amplitude of the current going through the antenna to continuously repeat a 128-bit signal. Each chip has a unique identifying number that links to a database. A cap made from special plastic covers half the capsule. The plastic is designed to bond with human tissue and stop the capsule from touring around your innards once it has been implanted. Oh yeah! If you have one of these inside of you and leave the company that had it implanted, then you will have to have an operation to remove it. An American security company that injected the tags into some of its workers went bust. That raises the question of who will pay to have the tag removed.

The tagging of human beings is on the increase. The proponents of tagging point to the perceived benefits of the technology; it can hold or link to information about the identity, physiology, health, nationality and security clearance of the person who carries

the embedded chip. According to an American company, VeriChip Corp, a subsidiary of Applied Digital Solutions, as of 2007, 2000 people have had chips implanted. The go-ahead for the tagging of human beings for medical reasons was given by the US Food and Drug Administration in 2004. In 2008 there was some discussion within the British Government about the possibility of implanting chips into prisoners. Following protests from Liberty, the idea seems to have been dropped. In the States the external tagging of prisoners is fairly widespread. It enables prison authorities to know the location of a prisoner 24/7. And in 2004 the Attorney General of Mexico and 18 of his staff had chips implanted to gain access to high security areas. If you are a gambler there are casinos and nightclubs that will chip you; the facility is on offer in Holland, Scotland, Spain and the good old US of A. By the time you get to the bar they'll have your favourite drink ready for you. VeriChip is pushing the technology to the American State machine; it is proposing a scheme for the tagging of military personnel as an alternative to metal dog tags. And its CEO, Scott Silverman, has proposed the chipping of guest workers entering the United States to assist the government in identifying them. Shortly afterwards, Associated Press quoted President Alvaro Uribe of Columbia as telling a US senator that he would agree to require Columbian citizens to be implanted

Your work uniform could be telling tales on you this very minute

with RFID tags before they could gain entry into the United States for seasonal work.[13] And what of the workers who would have to face the prospect of being chipped? Would they have a choice, or would that old slave-driver poverty take choice away from them? And who would pay for the implanting?

RFID tags being implanted into people is the eye-catching use of the technology, but there are less obvious applications being used to track us. As I mentioned above, your work uniform may be spying on you. The process involves the weaving of metallic fibres into work clothes so that your jacket becomes the tag. They can also put tags into work boots. That plastic ID card you have hanging from a lariat around your neck may have more uses than just telling people who you are. The same goes for the swipe card that allows you access to your workplace. Both items could be carrying information about you that could be linked to various databases; it could contain your National Insurance number, your timekeeping record, your disciplinary record and union affiliation. The latter could certainly be possible in workplaces where the company deducts union dues from the wage packet. They can be used to track your movements around a workplace, thereby policing any union activity you might be undertaking. And if information is

taken off your card you could find yourself locked out of the workplace. A few years back, when Ford sold the Halewood car plant to the Indian company, Tata, a dozen workers were locked out on the first day of work under the new owners. Their swipe cards wouldn't let them gain entry through the turnstiles. It is reckoned that they were identified as activists during the takeover negotiations between the two companies, and were thus surplus to requirements.

Computer monitoring, telephone management systems, hidden cameras and intelligent ID badges plus GPS and RFID are means to spy on us and keep us in line, but they are not the only means of surveillance used. A formidable array of other methods are also employed to gather information about us. They include: drug tests, background checks, intelligence tests, medical examinations and psychological tests (sometimes called psychometric testing). Companies will also visit sites like Facebook to try and discover what you are really like, as opposed to how you sell yourself at a job interview. A psychometric test paper I saw contained the following: 'You are working as a supervisor in a sheltered housing project and you discover that one of the residents is selling drugs. What do you do? A. Ignore it. B. Report it to your supervisor. C. Have a word with the person involved. D. Inform the police.' Well, folks, you are out of work and you want this job, so what box would you tick?

The 'War on Terror' has been used as the reason to increase the surveillance of the general population. Terrorists live and work in the communities they are prepared to attack, ergo, spy on the communities. The state's capacities for surveillance have been increased by the use of such systems as Echelon, a secretive project involving the intelligence agencies of the United States and other governments. Echelon monitors the global electronic spectrum, including telephone, email and satellite communications.[14] For the state, security overrides everything, so everything can be made subject to state surveillance – as those who participated in the riots during August will find out when their phone records are handed over to the 'Feds'.

As I'm finishing this article, the BBC news is telling me that over a hundred people have been arrested in Liverpool for involvement in the riots. Figures for London, Birmingham and Manchester are much higher. Surveillance cameras, phone records and photos from people's mobile phones have been used to gather the evidence for prosecution. And those prosecutions are being rushed through with indecent haste. As the welfare state model of social control is being dismantled, the need for other forms of control increases. The hegemonic structures are there to encourage us to interiorise the control mechanisms – the prison, the factory, the asylum and the school, for example. As sketched out in this article, capital and the state are using technologies like GPS and RFID to back up the already existing mechanisms of control.

The 'police' appears as an administration heading the state, together with the judiciary, the army, and the exchequer. True. Yet in fact, it embraces everything else. Turquet says so: 'It branches out into all of the people's conditions, everything they do or undertake. Its field comprises the judiciary, finance, and the army.' The police includes everything.[15]

We have to resist the internal and external policing of our lives.

Brian Ashton <brian.ashton00@gmail.com> is an ex car-industry shop steward who developed an interest in the logistics industry while doing support work with the sacked Liverpool dockers in the mid-'90s. He is currently researching the global supply chains of the clothing industry

FOOTNOTES

1 J. Markoff, 'Growing Compatibility Issues: Computers and User Privacy', *New York Times*, 3 March, 1999.
2 *Anarchy*, Vol. 7 No. 6, June, 1967.
3 Ibid.
4 Ibid.
5 Lyndall Urwick and E.F.L. Brech, *The Making of Scientific Management*, Vol.2, London, 1957, pp.10-11 quoted in Harry Braverman, *Labor and Monopoly Capital*, New York: Monthly Review Press, 1974, p.68.
6 Huw Beynon, *Working for Ford*, Allen Lane, London, Penguin Education, 1975, pp.22-23.
7 Braverman, op. cit. p.86.
8 Spector CNE, Case Studies, http://www.spectorcne.com/casestudies.html
9 See: David Crossland, 'German Anti-Snooping Law 'Long Overdue'', *Der Speigel*, 17 February 2011, http://www.spiegel.de/international/germany/0,1518,608223,00.html
10 Henry Chu, 'In Germany, widespread spying is back, this time by corporations', *Los Angeles Times*, May 23, 2009.
11 NLRB, Office of the General Counsel, Advice Memorandum, 26 February, 2003.
12 And it's still going on. See Angus Batey, 'Welcome to Royston … you're under surveillance', *The Guardian*, 29 July, 2011.
13 Kenneth R. Foster and Jan Jaeger, 'The murky ethics of implanted chips', http://spectrum.ieee.org/computing/hardware/rfid-inside/2, March 2007.
14 Michael Hardt and Antonio Negri, *Multitude*, London: Penguin Books, 2006, pp.202-203.
15 Michel Foucault, '"Omnes et Singulatim": Toward a Critique of Political Reason', in *Power*, J.D. Faubion (Ed.), New York: New Press, 2000, pp.318-319.

ANXIOUS RESILIENCE

Anxious subjects are also docile and self-absorbed subjects. The neoliberal state's production of generalised anxiety through non-stop risk pre-emption and contingency planning produces subjects that fit perfectly with the needs of capital – writes **MARK NEOCLEOUS.** *This is the concluding article to this issue's short series on surveillance, national security and logistics driven production*

I sit in one of the dives
On Fifty-Second Street
Uncertain and afraid

– W.H. Auden, 'September 1, 1939'

The bombs, bullets and gas that were to be launched across the world would prove that in September 1939 W.H. Auden was right to be anxious, but his uncertainty and fear never quite left him. In 1947 he published what was to be his final book length poem. It follows the lives of four characters, beginning as a conversation between them in a bar, and goes on to explore personal issues and western culture in the context of the defeat of Nazism and the rise of the Cold War. In so doing it unravels the ways that everybody these days 'is reduced to the anxious status of a shady character or a displaced person'. It is called *The Age of Anxiety*.[1]

Auden was hardly alone in thinking about the contemporary moment as an age of anxiety – in 1949 one Cold Warrior, Arthur Schlesinger, would declare in *The Vital Center* that 'anxiety is the official emotion of our time', and in 1950 Rollo May would publish *The Meaning of Anxiety* as a response to what he saw as a growing post-war condition. But the republication of Auden's book in the Spring of 2011 might make us wonder: why now? Why republish a 1947 book-length poem on the 'age of anxiety' in times which, we are told, are so very different from the years after WWII?

The republication of Auden's book could possibly be an attempt by the publishers to tap into what has become one of the central cultural

tropes of our time. In 1996, Sarah Dunant and Roy Porter edited a collection of essays on the 'age of anxiety' and, since then, the idea of such an age has become part of our cultural common sense, being used to think through questions of crime (*Fear of Crime: Critical Voices in an Age of Anxiety*, 2008); conspiracy theory (*The Age of Anxiety: Conspiracy Theory and the Human Sciences*, 2001); corporate management (*Global Firms and Emerging Markets in an Age of Anxiety*, 2004); parenting (*Perfect Madness: Motherhood in the Age of Anxiety*, 2005; *Worried all the Time: Overparenting in an Age of Anxiety and How to Stop It*, 2003); religions of all sorts (*Hope Against Darkness: The Transforming Vision of St. Francis in an Age of Anxiety*, 2002; *For Our Age of Anxiety: Sermons from the Sermon on the Mount*, 2009; *Ancient Wisdom for an Age of Anxiety*, 2007); language (*At War with Diversity: US Language Policy in an Age of Anxiety*, 2000); drugs (*The Age of Anxiety: A History of America's Turbulent Affair with Tranquilizers*, 2009; *A Social History of the Minor Tranquilizers: The Quest for Small Comfort in the Age of Anxiety*, 1991); new age claptrap (*The Road Less Travelled: Spiritual Growth in an Age of Anxiety*, 1997); sex (*Mindblowing Sex in the Real World: Hot Tips for Doing It in the Age of Anxiety*, 1995); food and drink (*Consuming Passions: Cooking and Eating in an Age of Anxiety*, 1998); and just plain old hope (*Hope in the Age of Anxiety*, 2009). This list could go on, but mention must be made of Haynes Johnson's history of national security drawing parallels between the McCarthyism of the post-war security state and more recent forms of McCarthyism being finessed in the War on Terror, called *The Age of Anxiety*.

This huge intellectual production parallels

developments in the psychiatric field. In 2013 a publication is due to appear called *DSM-V*. 'DSM' stands for the Diagnostic and Statistical Manual of Mental Disorders, and is the American Psychiatric Association's list of mental disorders and how to diagnose them. It is the essential text

---

On the basis of the DSM it might actually be impossible to be human and avoid being diagnosed with a treatable mental disorder connected with anxiety

---

for mainstream medicine, psychiatric practice and medical education across the globe. This one is called *DSM-V* because it's the fifth edition. The first edition in 1952 ran to 129 pages and contained just 106 diagnostic 'disorders'. Note: 129 pages, with 106 diagnostic categories. The second edition was published in 1968, with 134 pages and 182 categories. *DSM-III*, in 1980, was 494 pages long and contained 265 categories. *DSM-IV*, from 1994, had 886 pages and 297 diagnostic categories. *DSM-V* will be even larger and more substantial. Part of the increase in size and proliferation of categories has been because disorder has been defined according to forms of behaviour, so that aspects of our behaviour are used to define clinical categories. For example, being a bit nervous or shy is a symptom of an underlying condition, which then becomes a clinical category, such as social phobia, which is the term used as an explanation of what the manual calls 'social anxiety disorder' (SAD). Some of what it says about social anxiety concerns specific conditions, such as Parkinson's disease or disfigurement, but the term is also intended to capture fear or anxiety about one

or more social situations in which the person is exposed to scrutiny by others, such as having a conversation, being observed, performing; fear that one will act in a way that will be negatively evaluated; and fear of social situations which might provoke anxiety, and which are thus either avoided or endured with intense anxiety. *DSM-IV* then adds further detail on what it calls 'Generalized Anxiety Disorder' (GAD), which includes excessive anxiety and worry about two or more domains of activities or events such as family, health, finances, and school/work difficulties; excessive anxiety on more days than not and for three months or more; anxiety showing symptoms such as restlessness, edginess, muscle tension; anxiety associated with behaviours such as avoidance of situations in which a negative outcome could occur, or marked time and effort preparing for situations in which a negative outcome could occur, or procrastination due to worries, or seeking reassurance due to worries. And on it goes.

Note that the main way most of us would find ourselves in the pages of *DSM* – aside, that is, from sex (since, although homosexuality has been removed from the Manual, the inclusion of sex related diagnoses of all manner of desires, even having a low sex drive, means that it would not be difficult to find most of us in there somewhere) – is through the category of 'anxiety'. If one takes 'excessive anxiety' concerning two or more domains of activities or events such as family, health, finances and work, and one throws in some 'muscle tension' for good measure, it would be hard to find people who did not fit the category. On the basis of the *DSM* it might actually be impossible to be human and avoid being diagnosed with a treatable mental disorder connected with anxiety.

This would be consistent with the fact that, according to the World Health Organization,

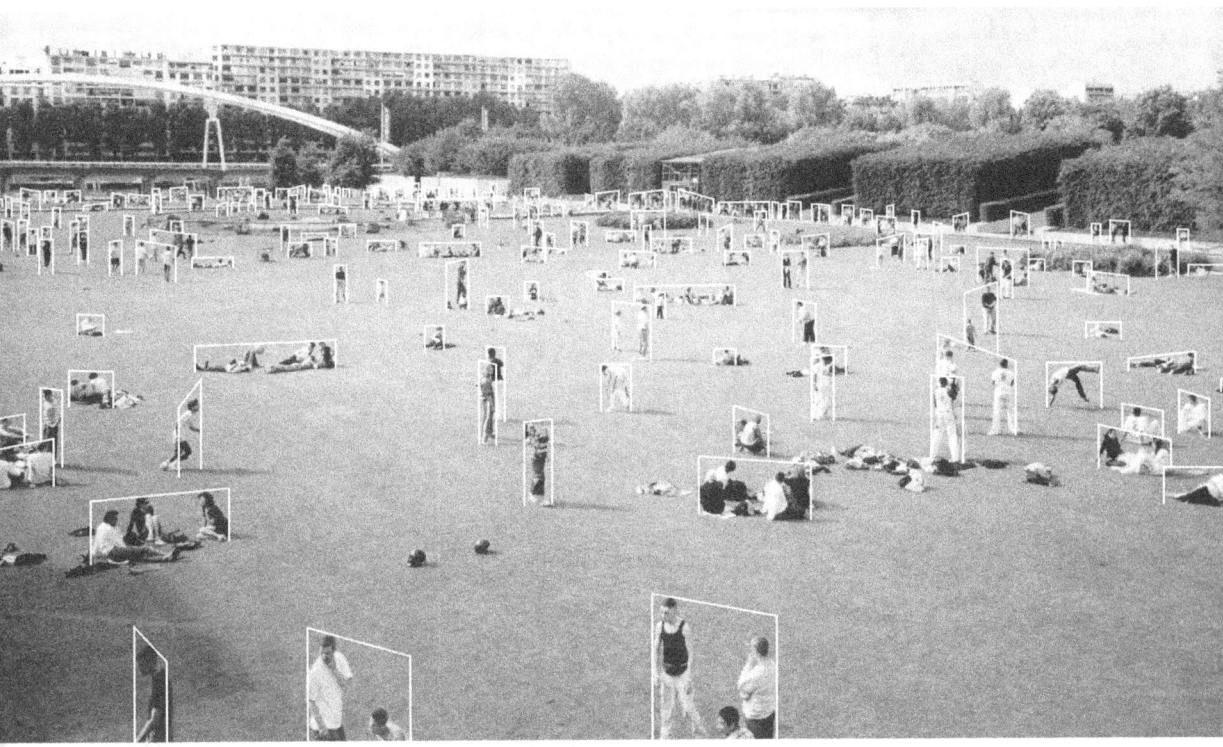

Isabelle Grosse, *Parc André Citroën*, colour photograph, 2002

anxiety has emerged as the most prevalent mental health problem across the globe (a process encouraged by the drugs industry, helping to generate an 'anxiety market' for drugs such as Paxil and Prozac, and extensive media coverage of the most recent 'anxiety' over anything from terrorism and social status down to cancer causing vegetables).[2] Thus one finds anxiety articulated everywhere. The Agoraphobia Society started life in the UK over 30 years ago with a fairly specific remit. It later became the National Phobics Society, with the remit extended along the lines of its change of name. It has recently renamed itself Anxiety UK. Perhaps symptomatically, what used to be called hypochondria is now officially 'health anxiety'. Freud, in his 1917 *Introductory Lectures*

on Psychoanalysis, makes the point that anxiety is the thing about which neurotics complain most.[3] In the light of the definition of more or less all of us as anxious, to be a neurotic citizen now seems to be a civic requirement.[4]

In this regard we might pay heed to Franz Neumann's comment on the role of anxiety as one of the cornerstones of the political mobilisation of fear under fascism.[5] But Neumann was also sensitive to the ways in which anxiety could play a similar role in the formation of liberal political subjectivity, one which opened the door to authoritarian mobilisations and manoeuvres. Might not that be especially the case in an 'age of anxiety' which is also an age of neoliberal authoritarianism? And how might that be connected to the fact

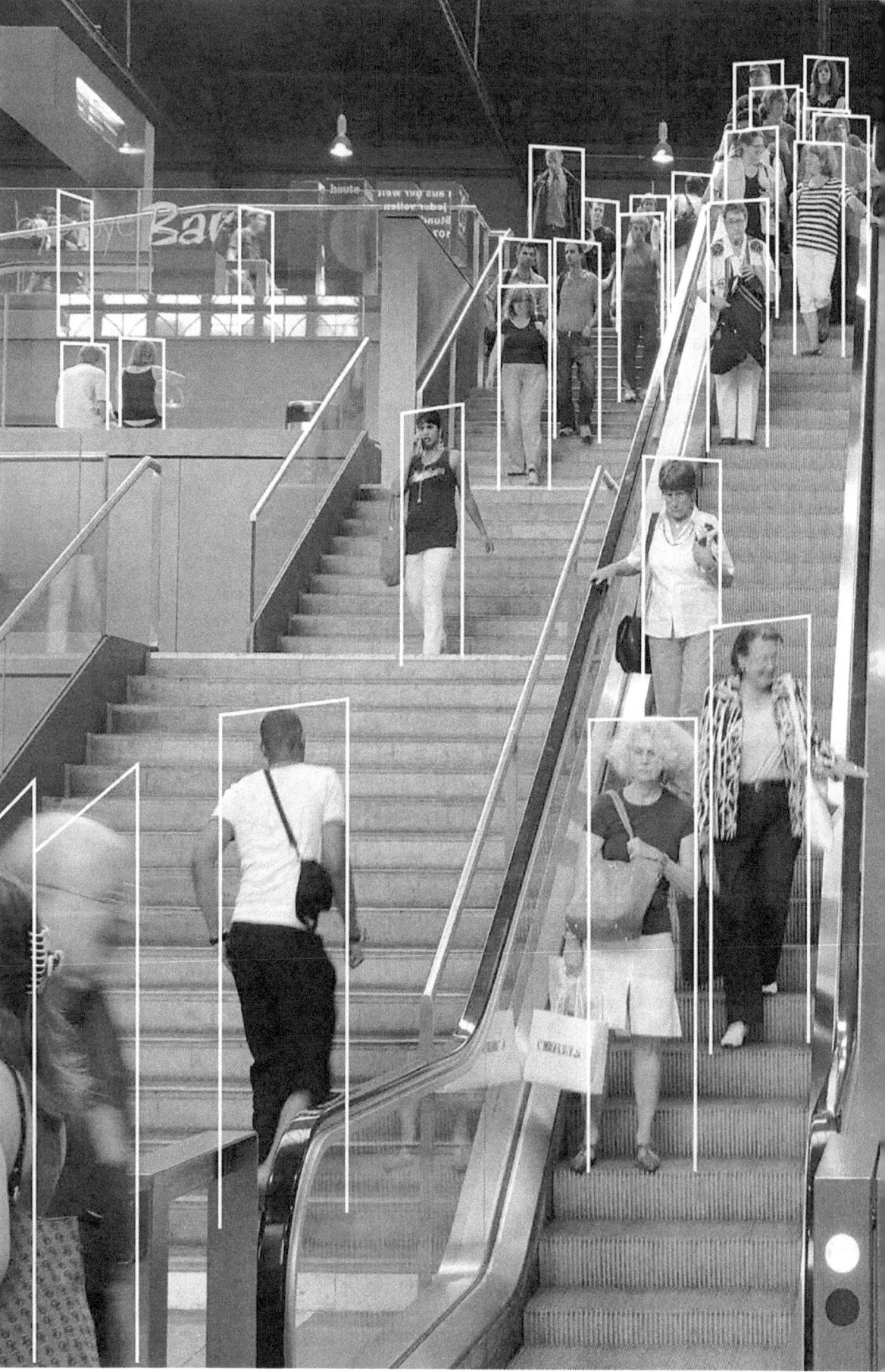

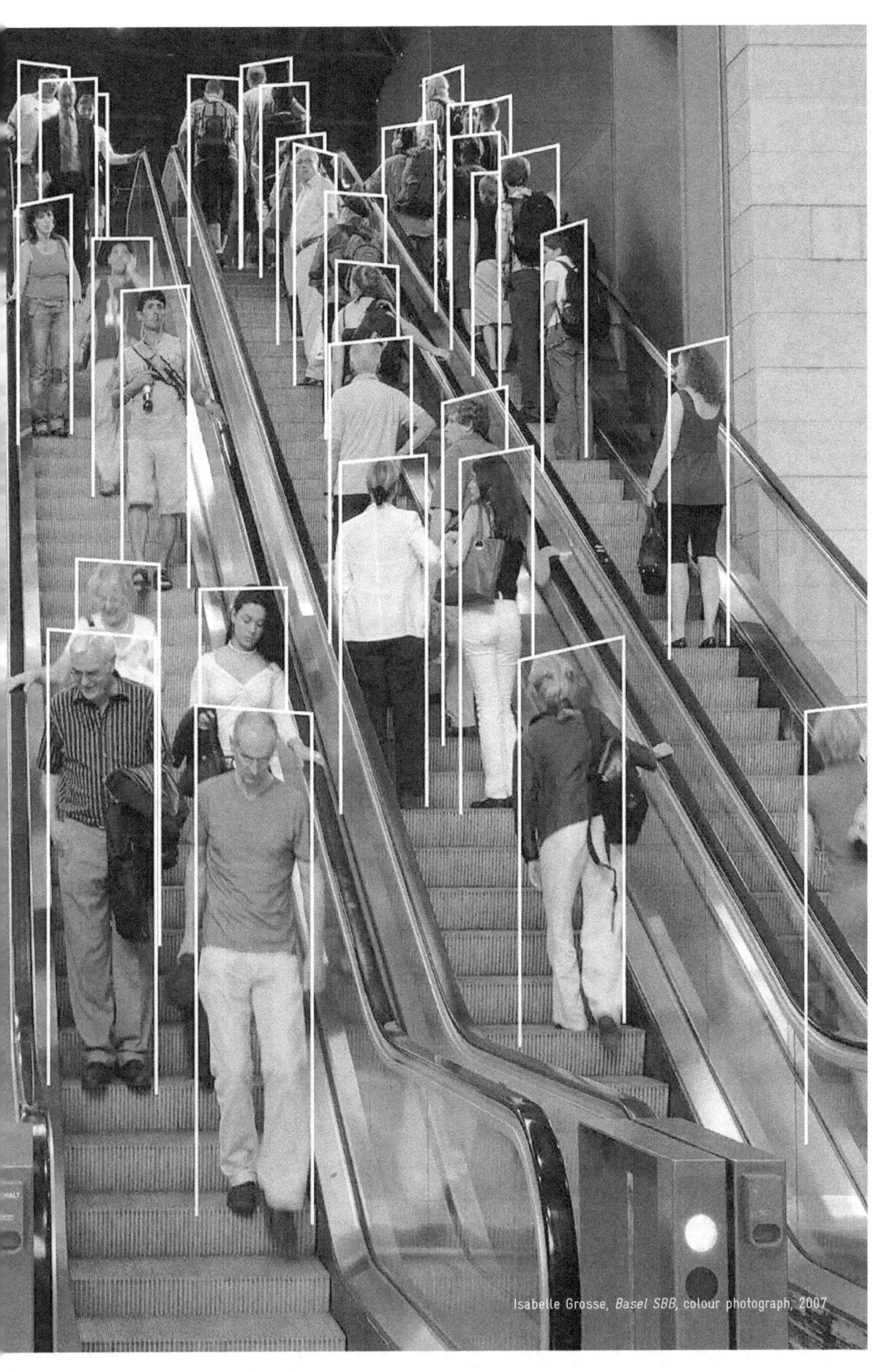

Isabelle Grosse, *Basel SBB*, colour photograph, 2007

that the age is also, if anything, an 'age of security'?

It is no exaggeration to say that the political production of subjectivity is now centrally driven by a security-anxiety field. A cursory glance at any security text, from the most mundane government pronouncement to the most sophisticated literature within 'security studies', reveals that through the politics of security runs a political imagination of fear and anxiety. I want to suggest that the management of anxiety has become a way of mediating the demands of security. It has done so within a broader logic of endless war. We have been told time and again that the War on Terror is a war like no other: this is a war without end, a permanent state of emergency, a peace which is also war. Because of this, the ideas of war and peace have been increasingly subsumed under the logic of police and security. Might we not think of the age of anxiety as a form of police power deployed for the security crisis of endless war?

One way to consider this is through the prediction of catastrophe and the anticipation of disaster that has come to the fore. A notable feature of political discourse has been the proliferation of ideas and categories centred on the idea that there is a disaster about to happen. Preparedness, prevention, planning and pre-emption have therefore become core ideas: everywhere one looks one finds emergency preparedness, contingency planning and pre-emptive action being addressed. Each of these is a concept with some scope, extending to war preparedness, disaster planning and terror attacks, and each of them resonates with and reinforces a whole gamut of associated security measures. They play heavily on the idea of potential 'natural disasters', but their real power lies in the presentation of endless war in terms of the coming political disaster. They are

intensely future oriented, in that they seek to shape behaviour towards a future event beyond our control, but which we must be prepared to take under our control. The worst case scenario must be prepared for, even though we don't know what it is yet and never can know what it is, and the preparations in question are a means of accommodating us to the security measures constantly established to deal with the catastrophe and disaster. Or, put differently: the security measures help us deal with the anxiety over the catastrophe to come. Anxiety has become a means of preparing us for the next attack in the permanent War on Terror. The attack, we are told time and again, is bound to come – how many times does a politician, police chief or security intellectual tell us that

anxiety is the thing about which neurotics complain most

'an attack is highly likely'? How many times are we told this even after a supposed victory in the war?[6] One which could be, and probably will be, worse than the last attack and might even be worse than anything we can imagine, all of which enables an acceptance of the ubiquity of the war, its claimed endlessness and the permanence of the security preparations carried out in its name.

Central to this process is the rise to prominence of the concept of 'resilience'. In the aftermath of the bombs in London in July 2005, Tony Blair spoke of 'the stoicism and resilience of the people of London', and Brian Paddick, then Deputy Assistant Commissioner of the Metropolitan Police, assured viewers that the emergency services 'had sufficient resilience to cope'. Their use of the term was significant.

'In the past few years', noted James Harkin in *The Times* in the aftermath of the bombing, 'the idea of resilience has been elevated to the most important buzzword in defence policy making circles. Since 11 September 2001, the Ministry of Defence has been busy commissioning all manner of research into the resilience of our big cities in the event of terrorist attack. Boffins in the Strategy Unit of No 10 have written countless turgid reports about what resilience means. Universities have set up whole departments, such as Cranfield University's Resilience Centre, to teach and study it.[7]

Resilience stems from the idea of a system (the term originates in ecological thought), and the official documentation on the term, of which there is now an enormous amount, plays on this: the 2008 OECD document on state building, styled 'from fragility to resilience', defines the latter as 'the ability to cope with changes in capacity, effectiveness or legitimacy. These changes can be driven by shocks... or through long-term erosions (or increases) in capacity, effectiveness or legitimacy'.[8] A key United Nations document on disaster management suggests that resilience requires 'a consideration of *almost every physical phenomenon on the planet*'.[9] Note: almost every physical phenomenon on the planet. Although the overall argument is couched in terms of physical risks, the UN links it explicitly to the wider security agenda in a way which connects it intellectually and politically to domestic legislation such as the UK's Civil Contingencies Act 2004, which involves contingency plans for anything which might be said to affect the 'welfare' of the UK. The extent to which 'resilience' has come to the fore in the politics of planning is witnessed by the London Resilience Team set up to 'deliver Olympic Resilience in London', and the extent to which it is designed

Might we not think of the age of anxiety as a form of police power deployed for the security crisis of endless war?

to connect emergency planning to the logic of security is evidenced in the fact that the London Organising Committee of the Olympic Games has a 'Security and Resilience' section.

As these examples suggest, in terms of state power, huge resources are now spent to map out potential disasters and apprehend coming disasters. Playing on the origins of resilience in systems thinking, the idea of planning out organisational and institutional resilience has become a key strategy across central state agencies, local governments, emergency services and health authorities. In the UK, for

capital thrives on anxiety

example, this would stretch from the creation of 'UK Resilience' based in the Cabinet Office right down to the fact that sniffer dogs, like their handlers, now 'have to be resilient'.[10] There has also developed a commercial rhetoric of 'organisational resilience' for corporations.

This rise of 'resilience' during neo-liberalism's development is significant. Although this connection might seem odd given that, more than anything, resilience assumes a massive state role in planning for the future, the point of this future is that it is unknown and uncertain. Thus, as a political category, resilience relies fundamentally on an anxious political psyche engaged in an endless war and perpetually preparing for the coming attack. Such a strategy foregrounds a politics of anticipation, in which the anticipation itself becomes both an exercise and a source of anxiety. But the term has been expanded to straddle the private as well as the public, the personal as well as the political, the subjective as well as the objective, and so *organisational* resilience is

connected to *personal* resilience in such a way that contemporary citizenship now has to be thought through 'the power of resilience'. Thus one finds a set of texts on personal resilience which would not be out of place were they situated on the same bookshelves as the works on anxiety cited above: texts about resilience as a personal attribute in which citizen-subjects are trained to 'achieve balance, confidence and personal strength' or, in the subtitle of another, 'find inner strength and overcome life's hurdles', or better still, just 'bounce back from whatever life throws at us'.[11] The anxious citizen is acknowledged as the resilient citizen and championed.

It is here that one finds the relationship between the economic development of neoliberal subjectivity and the political development of resilient citizenship. Marx long ago spelt out the ways in which capital, as a system rooted objectively in uncertainty, instability, restlessness, agitation and change, generates the very same subjective feelings in the workers it uses; capital thrives on anxiety. The neoliberal intensification of this process – repackaged by politicians and employers as an inevitable fact of contemporary labour and exacerbated by the anxiety associated with the rise of consumerism, a decline of trust in public institutions and private corporations, and a collapse in pension schemes – has been compounded by this articulation of resilience as something personal as well as systemic. Resilience is thus presented as a key way of *subjectively* working through the uncertainty and instability of contemporary capital. The neoliberal subject can 'achieve balance' across the several insecure and part-time jobs they have, can 'overcome life's hurdles' such as facing retirement without any pension to speak of, and just 'bounce back from whatever life throws at us', whether it be the collapse of welfare systems or global economic meltdown. The policing of the resilient citizen coincides with the socio-economic fabrication of resilient yet flexible labour. Neoliberal citizenship is nothing if not a *training in resilience*.

All of which is to say that anxiety and resilience are now core to the jargon of neoliberal authenticity.[12] Superficially, such jargon is full of 'recognition' for the complexities of human experience ('of course you are anxious'; 'we all share the same fears'; 'it's only natural to be anxious'), but this merely encourages the naturalisation of a neoliberal subjectivity mobilised for security and capital: the jargon of neoliberal authenticity is the jargon of neoliberal authoritarianism. This is police power at its most profound, shaping subjectivity and fabricating order through counselors within police departments, therapists within the community, psychologists in the media, and experts working in the cultural field, all offering advice on our anxieties and coaching us in our resilience. And it is a police power par excellence in closing down alternate possibilities: we can be anxious about what might happen, but our response must be resilience training, not political struggle. We can be collectively anxious and structurally resilient, but not mobilised politically.

Mark Neocleous <mark.neocleous@brunel.ac.uk> is Professor of the Critique of Political Economy at Brunel University, UK, and on the editorial collective of the journal *Radical Philosophy*. He is author of *Critique of Security* (2008), *The Fabrication of Social Order: A Critical Theory of Police Power* (2000), and a range of other books and articles. He has a co-edited book forthcoming called *Anti-Security* (2011). His current project is a work of counter-strategic theory organised around the concept of pacification

FOOTNOTES

1 W.H. Auden, *The Age of Anxiety*, Princeton: Princeton University Press, 1947; reissued 2011, p.3.

2 'The WHO World Mental Health Survey: Global Perspectives on the Epidemiology of Mental Disorders', Cambridge: Cambridge University Press, 2008.

3 Sigmund Freud, *Introductory Lectures on Psychoanalysis, Part III: General Theory of the Neuroses* (1917), in *The Standard Edition of the Complete Psychological Works of Sigmund Freud*, Vol. XVI, London: Vintage, 2001, p.392.

4 Engin F. Isin, 'The Neurotic Citizen', *Citizenship Studies*, Vol. 8, No. 3, 2004, pp.217-35.

5 Franz Neumann, 'Anxiety and Politics' (1954), in Neumann, *The Democratic and the Authoritarian State*, New York: Free Press, 1957.

6 Most recently, in this comment from Sir Paul Stephenson, Head of the Metropolitan Police, following Bin Laden's killing, May 2011, cited *The Daily Telegraph*, 5 May 2011.

7 James Harkin, 'What is Resilience?, *The Times*, 9 July, 2005.

8 'OECD, Concepts and Dilemmas of State Building in Fragile Situations: From Fragility to Resilience', OECD, 2008, p.17.

9 'United Nations, Living With Risk: A Global Review of Disaster Reduction Initiatives', Vol. 1, New York and Geneva: UN, 2004, p.37.

10 Police Inspector Alun Jenkins, cited in 'Sniffer Dogs Prepare for London Olympics', BBC News, 15 October, 2010. For 'UK Resilience' see http://www.cabinetoffice.gov.uk/ukresilience.

11 Robert Brooks and Sam Goldstein, *The Power of Resilience: Achieving Balance, Confidence, and Personal Strength in Your Life*, New York: McGraw-Hill, 2004; Karen Reivich and Andrew Shatté, *The Resilience Factor: 7 Keys to Finding Your Inner Strength and Overcoming Life's Hurdles*, New York: Broadway Books, 2003; Jane Clarke and John Nicholson, *Resilience: Bounce Back from Whatever Life Throws at You*, Richmond, Surrey: Crimson Publishing, 2010.

12 I am playing here on Theodor Adorno, *The Jargon of Authenticity* (1964), Knut Tarnowski and Frederic Will (Trans.), London: Routledge and Kegan Paul, 1973.

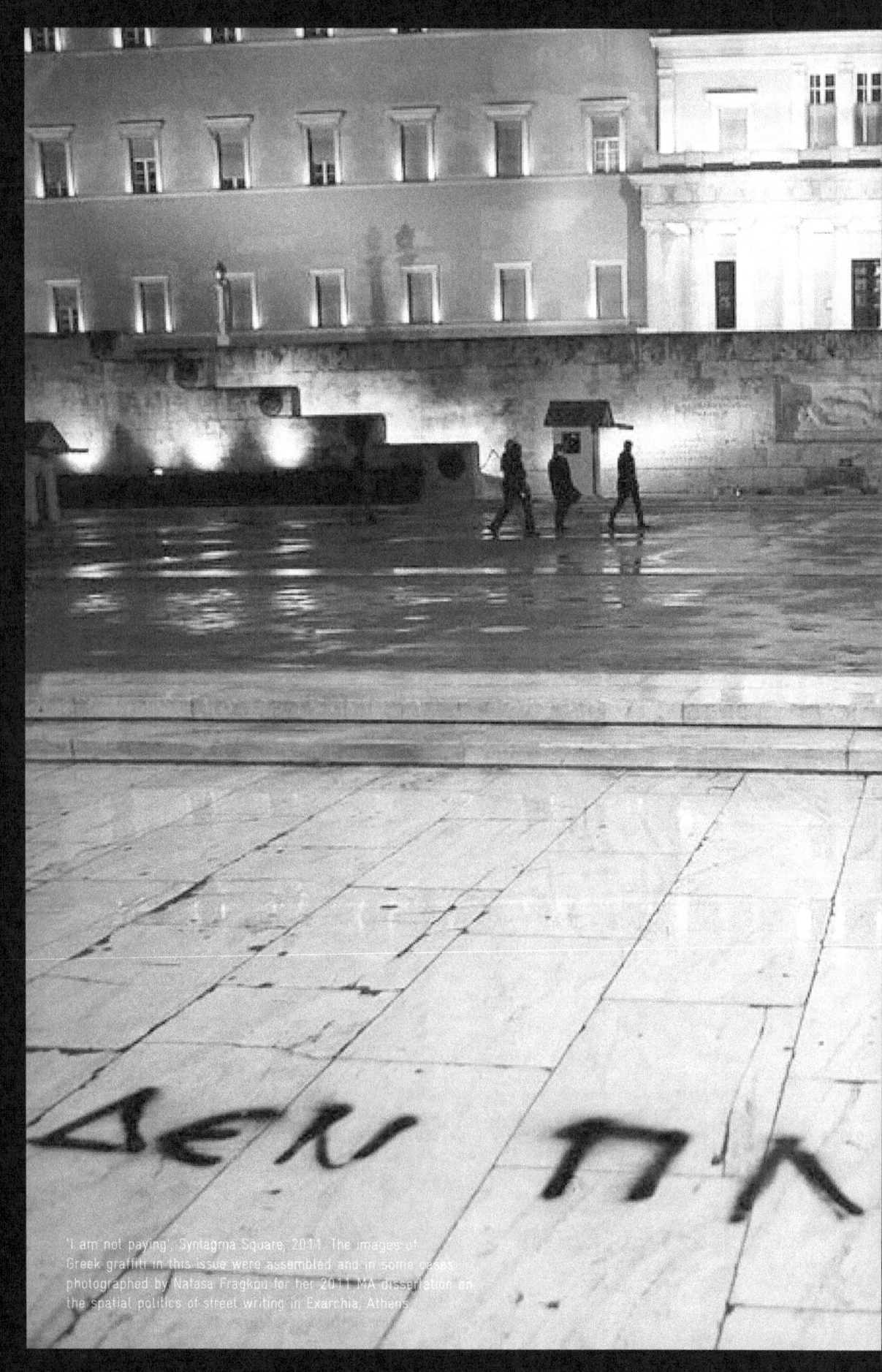

'I am not paying', Syntagma Square, 2011. The images of
Greek graffiti in this issue were assembled and in some cases
photographed by Natasa Fragkou for her 2011 MA dissertation on
the spatial politics of street writing in Exarchia, Athens.

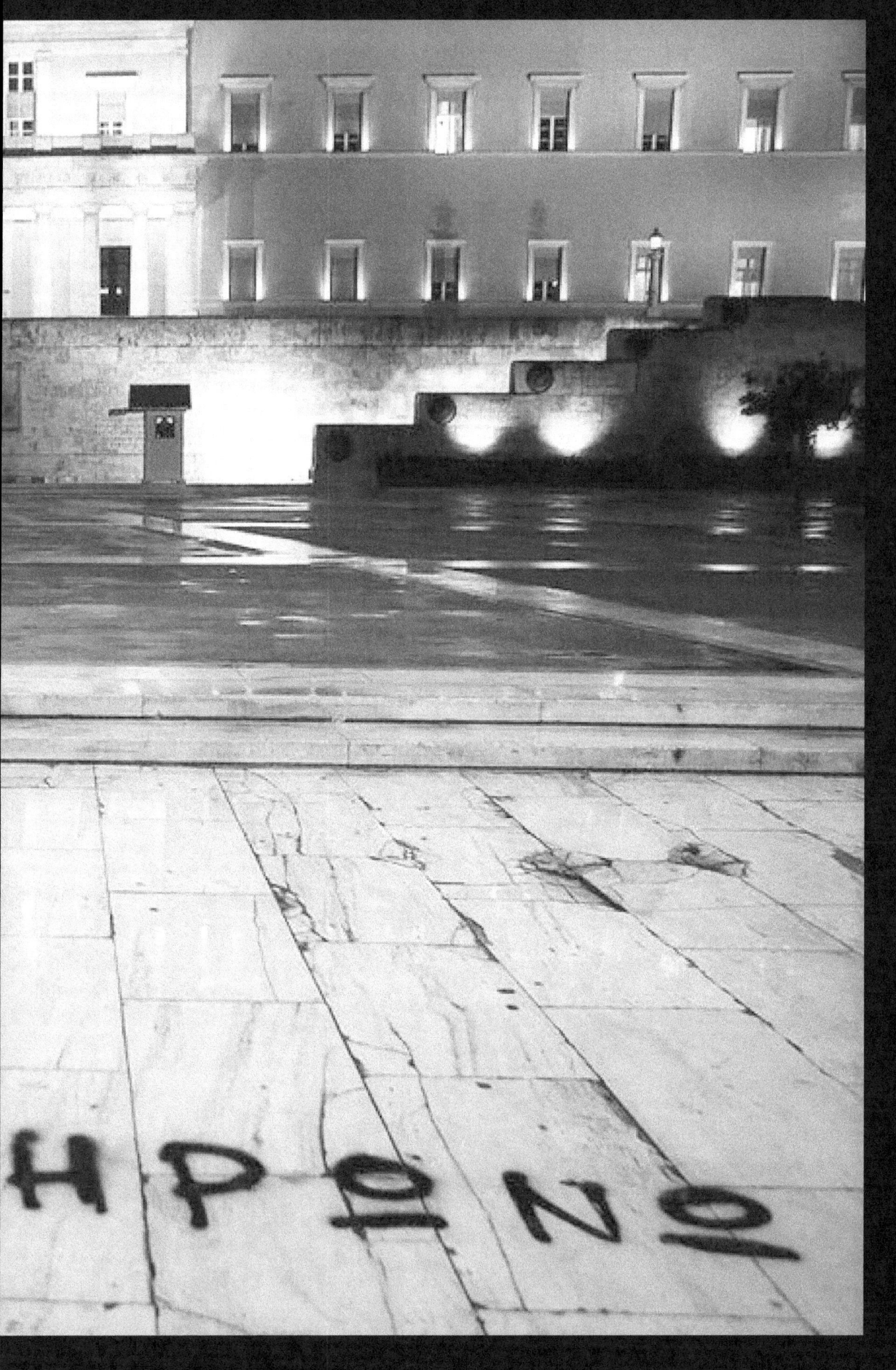

SPAGHETTI COMMUNISM?

If Westerns allegorise a mythical space of gradual resolution and order, the western all'italiana explodes the American dream of stabilising prosperity with excessive violence and explicit anti-colonial themes. BENJAMIN NOYS *argues for a deeper analysis of an intensely political cinematic genre*

CLEANING UP THE WHOLE WORLD

Gilberto Perez remarks that 'the Western doesn't just tell violent stories, it tells stories about the meaning, the management of violence, the establishment of social order and political authority'.[1] Perez elsewhere concedes that the Western runs 'a gamut of political persuasions',[2] but is keen to emphasise that in the classical American Western this 'management of violence' takes the form of a 'vital dialectic'[3] in which is synthesised a *civilized violence*.[4] Serving his deliberately provocative re-imagination of the 'frontier' as equivocal site of liberty, Perez regards the Western as the romance of the birth of a new political order through the, often literal, marriage of East and West, in which violence plays the role of a 'vanishing mediator'. Such an argument hardly seems to hold for the Italian Western of the 1960s and 1970s, often known affectionately or derogatorily as 'Spaghetti Westerns', in which the excessive hyper-violence associated with the form makes it difficult to see how it might be pressed into service for a 'vital dialectic' of 'civilised violence'. The very excess of the violence on display, as well as its displacement from the 'mythological' place of America, fragments any dialectical sublation of violence within a national or political order.

This suggests a very different 'political persuasion', and very different questions concerning the 'management of violence'. In fact, objections to Spaghetti Westerns, often by critics enamoured of classic American Westerns (or 'Hamburger Westerns'[5], in Christopher Frayling's mischievous suggestion), were usually founded on their 'excess' of violence. Philip French, writing in 1972, describes a filmography of continental Westerns as 'to me read[ing] like a brochure for a season in hell.'[6] A surprisingly apposite comment as we will see. Spaghetti Westerns, in fact, constructed a form of violence that carried a rather different and more intense charge. Franco Nero, who played the eponymous 'Django' in the seminal Spaghetti Western, remarked:

> Spaghetti Westerns were for a certain kind of audience – the workers, I think. Mainly workers, boys... yes, all kinds of workers – and the workers they fantasize a lot, and they would like to go to the boss in the office and be the hero and say 'Sir, from today, something's going to happen.' And then – bam, bam! they want to clean up the whole world.[7]

A rather extreme example of the refusal of work, although if one considers the strategies and intensity of conflict in Italy between 1968-1977 – 'Our Comrade P.38' as one anonymous tract had it – 'clean[ing] up the whole world', gains a prescient resonance.[8]

This is reinforced by Johanna Isaacson's argument that the genre films of the late '60s and '70s belong to a 'moment when it was taken for granted that genre film was political to the bone, reflecting the subjectivity, anger and tastes of a radicalized proletarian sensibility.'[9] The question of violence, in terms of audience, turns here on sensibility: bourgeois or proletarian? The Spaghetti Western is, I would argue, exactly the archetypal film form of this moment, to quote Isaacson again, 'appealing to both [the] proletarian desire for spectacle and for representations of political repression.'[10] Although this schema of divided sensibility is too simplistic, not least in its supposition of a unified 'proletarian sensibility', it draws attention to the 'class' charge of violence emergent in these films. While this often takes overtly and unequivocally political forms, as we will see, what I want to focus on here are a small number of films that take their 'representations of political repression' into the realms of what Gail Day, in a very different context, has identified as a 'left-oriented nihilism'.[11] These are Sergio Corbucci's *Django* (1966) and *The Big Silence* (1969) (also known as *The Great Silence*), and Guilio Questi's gothic horror Spaghetti Western *Django Kill! / If You Live Shoot* (1967). Produced and shown on the cusp of the eruption of the most militant workers' movement in Europe, these films display a striking nihilist politics that internalises and prefigures the experience of defeat.

POPULAR EXCESSIVE VIOLENCE

First, some context: between 400 and 450 Italian westerns were made, according to Christopher Frayling, in the period from 1963 to the mid-1970s.[12] The most familiar are obviously the works of Sergio Leone, who broke out from the 'ghetto' of popular filmmaking into the category of auteur. The 'other Sergio' – Sergio Corbucci (1927-1990) – is perhaps a more representative figure of the cycle, especially with his work *Django*. It should be noted that although Spaghetti Westerns are often regarded as hyper-violent works, a large number were 'guns and gadgets' Westerns, heavily indebted to the Bond movies and with a comic streak, such as the charming *Sabata* (1969), starring Lee Van Cleef complete with four-barrelled derringer.

Broadly to characterise the whole 'cycle' of Italian Westerns, we can borrow Philip French's comment on post-*Wild Bunch* American Westerns:

> At a social level the movies are reflecting current concerns and anxieties; from a commercial point of view a profitable subject is being exploited that seems to go down well at the box office; viewed aesthetically, the cycle of movies is offering a cumulative series of variations upon an established theme.[13]

In terms of the aesthetic 'variations' it is worth noting that many of the instances that seem most singular to the Italian Western, especially of masochistic violence, in fact occurred previously in American Westerns or in the immediate source material for Spaghetti Westerns: Kurosawa's *Yojimbo* (1961). *Yojimbo*, its plot almost certainly derived from Dashiell Hammett's novels *The Glass Key* (1931) and *Red Harvest* (1929) (Hammett was an anti-fascist who joined the American Communist Party in 1937, pleaded the Fifth in a case linked to the communist witch hunts in 1951, served time in prison for contempt of court, and was later blacklisted), literally set the pattern of the lone hero playing off two gangs against each other to their mutual destruction, and also the tendency to quasi-homosocial or homoerotic torture scenes.

So, we are talking here of what Christopher Frayling calls 'formula cinema', but at the same time we have to recognise that this was an intensely *political* genre cinema.[14] Obviously, as I've just noted, its source material is broadly left-wing, with Hammett's account of corruption and collusion linking to the general 'populist' politics of the Western (although we should well note, as Philip French does, the limits of that 'politics': 'the Western is ill-equipped to confront complex political issues in a direct

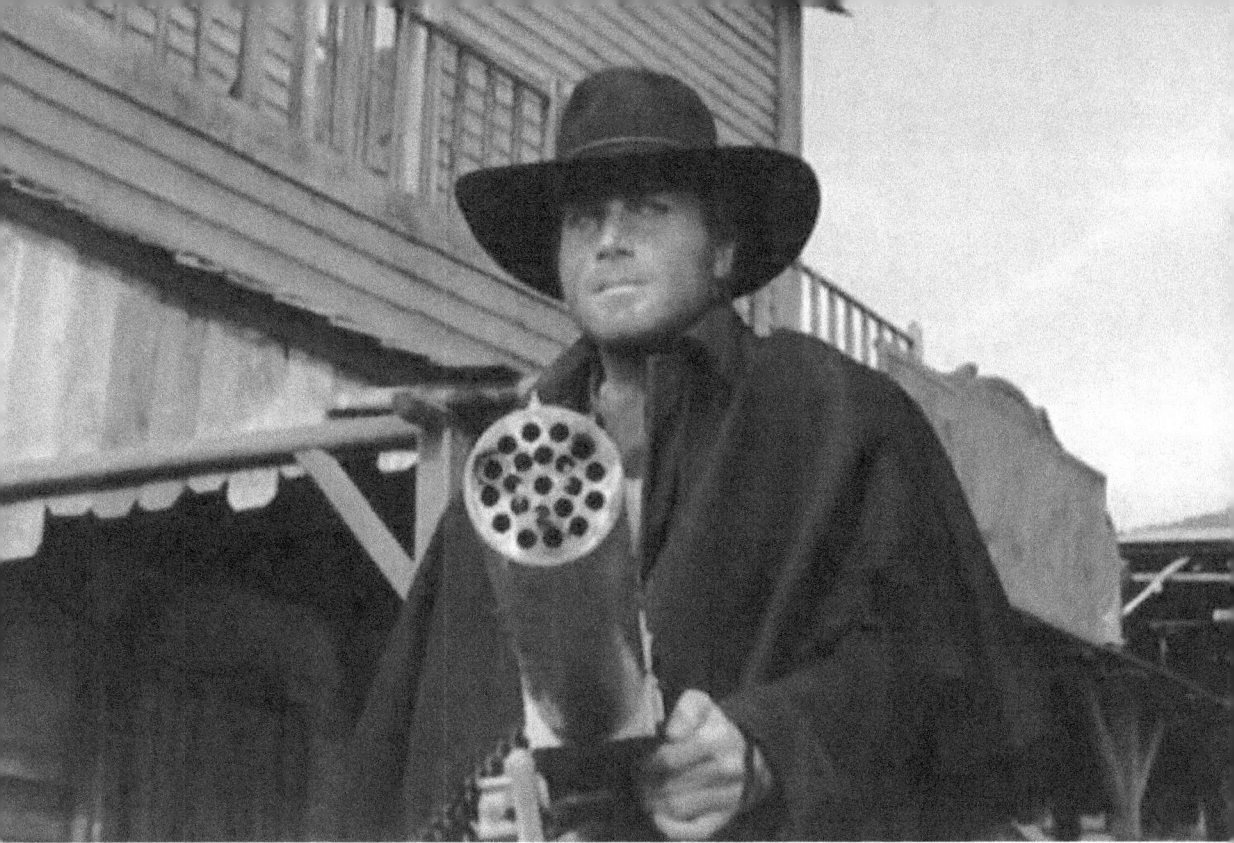

Franco Nero in Sergio Corbucci's *Django* 1966

fashion. The genre belongs to the American populist tradition which sees all politics and politicians as corrupt and fraudulent'.[15]) Also, to court the 'intentional fallacy', many of the directors and writers of these films were men, and yes men, of the left; either communists or sympathisers often energised by the emergent struggles of the 1960s, especially the Cuban revolution and the struggle of the Vietnamese against the Americans.

In the extensive debate on the politics of the possibilities of 'popular' film versus more Brechtian and modernist strategies of alienation that took place in the late 1960s and early 1970s, it may be surprising now to realise that Spaghetti Westerns played a key role. Pierre Baudry, writing in *Cahiers du Cinema* during its most *haut*-Marxist period, noted, in 1971, the shifting and recurring patterns of these genre films, especially in their exploration of the dynamics of colonialism and revolution through moving from the 'Gringo'/Bandit pairing to the 'Gringo'/Mexican revolutionary pairing. Ultimately he found wanting this 'commercial' cinema, preferring the austere path that was

to be taken by Jean-Luc Godard and Jean-Pierre Gorin's Brechtian critique of the Western, *Vent D'Est* (1970).

In fact, much of the political discussion of the Spaghetti Western has focused on these 'pairing' films, which contain obvious reflections on Vietnam, as well as Italy's own situation. The best of these is probably *A Bullet for the General* (1967), which was scripted partly by Franco Solinas, who was also responsible for writing *The Battle of Algiers* (1966), and for the script for what we could consider as the finest film on this theme of coloniser/colonised pairings: *Queimada / Burn!* (1969), which were both directed by Gillo Pontecorvo. Solinas' impeccable political credentials, his deliberate decision to work in the popular medium of the Western as a political act, and his sophisticated inclusion of Fanonian themes, all make the politics of these films striking and evident. What I am concerned with are films with a rather less direct politics, a politics in which the excess of violence is not placed in a 'revolutionary' or anti-colonial context, but operates in a more 'free-floating' and ambiguous form.

EPIC NIHILISM

Key to my analysis is the conjugation of 'epic nihilism', derived from Badiou's analysis in *The Century* where he remarks: 'may your force be nihilistic, but your form epic.'[16] We can find this conjunction already encoded in the Ur-work of the Spaghetti Western genre: Sergio Corbucci's *Django*. Here, the epic form of the Western is, literally, dragged through the mud – in its striking opening sequence in which Django drags a coffin through the mud; a coffin, as we later find out, that contains the machine gun with which he will exterminate his adversaries. The town at the centre of the usual plot of playing off rival gangs is bathed in mud, and the film ends in a gunfight in a cemetery in which Django, with smashed hands, painfully and finally manages to shoot his chief tormentors after mounting his gun on a grave cross. It is not difficult to identify the mud as an allegory of the practico-inert, with Django becoming mired in the inertia that seems to afflict the supposedly decisive Spaghetti Western hero.[17]

This is reflected in the constant *delay* of revenge which affects Django, and many of the other heroes of these films. Once they become involved in double-crossing the competing gangs, these heroes persistently fail to act and as a result are usually tortured before exacting 'final' revenge. The films themselves, despite their bursts of hyper-violent action, are also tortured in their following of this repetitive path of delay and finally action. Of course, we could make the usual references to *Hamlet*, or to psychoanalytic explanations based on the displacement of murderous desires, but it strikes me how these films also mimic the affective texture of the working day. Django's own 'mechanical' killing, carried out with the Gatling gun, is over promptly, but seems to

leave him as mere 'appendage' of the machine (to use Marx's phrase). The freelance 'labour' of the gunfighter is filled with longueurs, in a state of a kind of proto-precarity awaiting a new contract, or failing to execute a supposedly personal and pressing desire. What we have here is a strange tempo of labour that retards and confines action to sporadic outbursts of 'acting out', which appears to require the extremities of torture to 'activate'. Even the recurrent trope of the usually deliberately inflicted injury to the hero's gun hand, which can be found in *Django* and other films, seems to have the echo of the industrial accident. Despite the 'hopeful' ending of *Django*, in which the hero escapes with his life, his time as a gunslinger is presumably over.

In fact I wonder if these films do not take place in the 'factory-universe' described by Maurice Blanchot.[18] This is a space of infinite repetition, excess, and the vacancy of Being. The deliberately hellish towns which our heroes tarry in figure this space. As Blanchot puts it, of the factory:

> There is no more outside - you think you're getting out? You're not getting out. Night, day, there's no difference, and you have to know that retirement at sixty and death at seventy will not liberate you. Great lengths of time, the flash of an instant - both are equally lost.[19]

The factory is the space of infinite excess, of 'the infinite in pieces', figured in the broken and ruptural spaces through which the Spaghetti Western hero drifts, or becomes mired.[20] These enclosed towns are not the wide open plains or vistas of monument valley, or even Almería, but have no more outside; are circles of hell (a metaphor literalised by Clint Eastwood in his post-Spaghetti Western *High Plains Drifter* (1973), as his hero has the town road sign painted red and renamed 'Hell').

INTERMINABLE INERTIA

In Corbucci's later *The Big Silence* it will be snow that performs a similar function of signifying this inertial time. These two films by Corbucci are, to borrow Maurice Blanchot's phrase, 'condensed around thick living substances, which are at once over-abundantly active and of an interminable inertia.'[21] We can take this as a certain coagulation of living labour in dead labour, and dead time, in which the performance of virtuosity is only ever fleeting, and forever punished. The 'production line' of killing runs on receding amounts of living labour, as value production is mechanised into the machine gun. Taking this motif of inertia to the extreme, *The Big Silence* also takes the usual taciturn Western hero to the limit, with the character of 'Silence' (played by Jean-Louis Trintignant), who is mute due to mutilation by bounty hunters (or 'bounty killers' as the film usually prefers) when he was a child. Again, we have the tempo of freelance labour, as Silence intervenes in a small-scale war between the 'bounty killers', who 'operate according to the law', in pursuing the former townspeople and farmers who have been driven into banditry by the actions of Pollicutt, the banker and Justice of the Peace.

This political fable, which follows closely the usual script of political populism – good people driven to 'social banditry' by a corrupt law – is complicated by the film's own seeming lack of faith in this story. Silence works for money, but works, again, in a lackadaisical and intermittent fashion. In contrast, the leading bounty killer Tigrero (an excellent performance by Klaus Kinski), is a model of sadistic efficiency: killing in the most expedient fashion, loading his dead victims onto the stage, and assiduously collecting his 'reward' (with a cut going to the banker Pollicutt). Upbraided by the new sheriff, Tigrero remarks: 'Every business has its own risks and rules'. Later, after having got the drop on the sheriff, who has lectured him on justice replacing violence, Tigrero kills him and remarks the only law is 'survival of the fittest': *Homo homini lupus*, although, as Tigrero notes to his friends 'when are wolves afraid of wolves?'

Of course the true destruction of this fable of populism, and proof of the power of the 'representation of political repression', is the film's ending. The comedic sheriff character is drowned in a frozen river when Tigrero shoots out the ice from under him to ensure an 'accidental' death. The town's prostitute matron, who had a touchingly comic and halting relationship with the sheriff, is shot by Tigrero after he has baited her with news of the sheriff's death. Although the sheriff had planned to feed the 'bandits' pending an amnesty this plan now turns into a fatal trap as Tigrero's men capture them when they come for the food. Silence, his hands ruined in a fight, and his female companion, wife of one of the men he is avenging, are gunned down by Tigrero after Silence refuses to flee and chooses instead to fight. As a result the 'bandits', tied-up in the saloon, are massacred. The 'civilising vital dialectic' of violence is broken, but, as Tigrero says, 'all according to the law'. He and the bounty killers plan to return to collect their now considerable bounties, as the distinction of law-making and law-preserving violence is broken through the 'law' of original accumulation that pays for the necessary violence required at all points.

FOUL GOLD

These thematics reach their baroque extreme in Guilio Questi's *Django Kill!*. Questi was not interested in making a Western. Instead, when

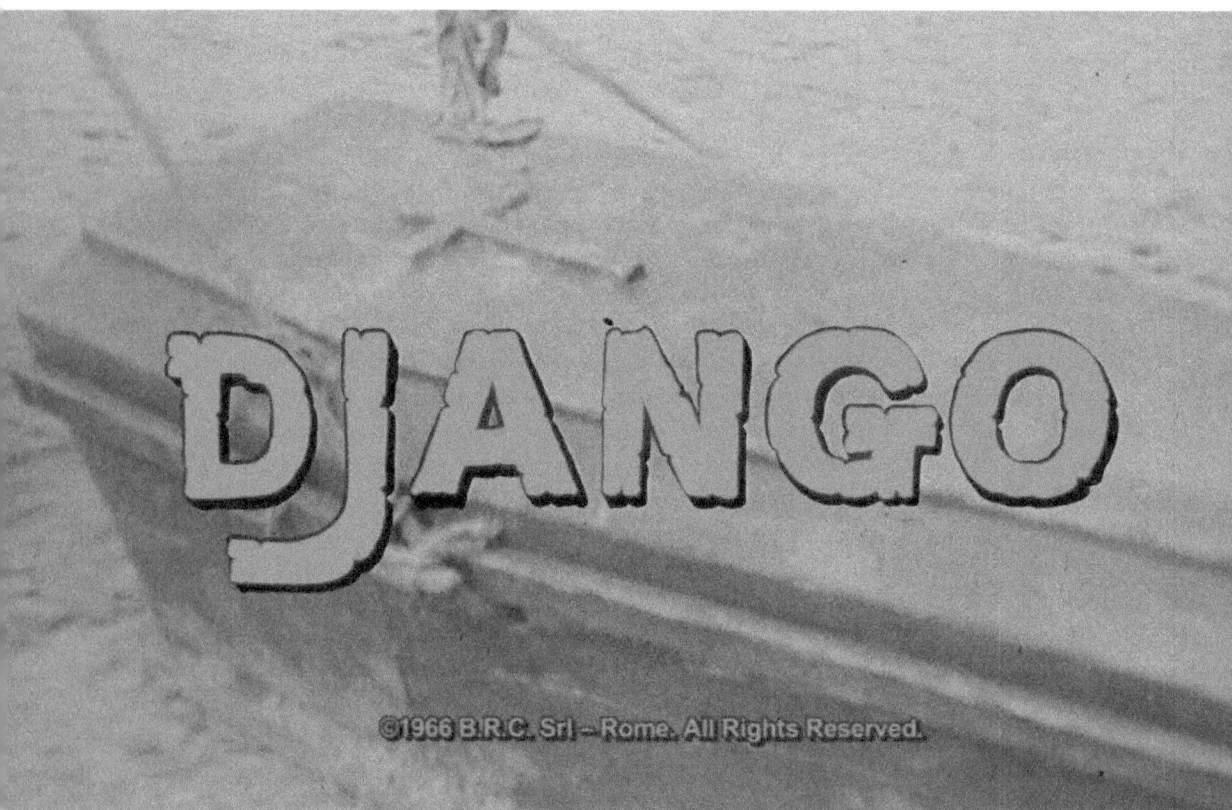

Opening credits, Sergio Corbucci's *Django*, 1966

offered such a project he took the opportunity to make a more personal film that dealt with his experiences as an anti-fascist partisan: 'I wanted to recount all of the things, the cruelty, the comradeship with friends, the death, all the experiences I had of war, in combat, in the mountains.'[22] The result is a work of convulsive and violent beauty. If Jansco's *The Round-Up* (1965) is a film of the balletic choreography of physical repression, *Django Kill!* is a film of violence, sexual and physical, as carnivalesque, and the non-sequiturs of repressive desublimation.

It begins with the hand of the central character, the stranger (played by Tomas Milian), emerging from a grave to a surprisingly jaunty Western tune. In a series of bizarrely edited flashbacks (at one point a body appears to roll uphill in a reverse of the actual shot), we learn he was betrayed by a gang of outlaws led by the racist Oaks after their successful robbery of a gold shipment. Rescued by highly unlikely mystical-hippy 'Indians', who smelt his share of the gold into bullets, the stranger determines to take revenge on the gang.

The outlaws, meanwhile, have arrived in quite the most disturbing town, which makes *Dogville* look like a good choice for a holiday, and is known by the Indians as 'the unhappy place'. Riding in they see a naked boy playing with himself, a girl twisting the hair of a playmate, a man retching, a young girl under the boot of 'uncle Max', a woman threatening to bite her husband, and a crippled hedgehog (!). Soon recognised by the townspeople, the outlaws are killed in a carnivalesque episode of 'civilising' violence; complete with beatings, hangings, stringing-up bodies, drowning, and close-up head shot executions. Arriving in time to find Oaks holed up in a store and fighting for his life, the stranger agrees to take $500 for killing him. Confronting Oaks, who remarks, 'you've

come back from hell', they engage in a quasi-comedic shoot out. Oaks is left bullet-ridden but still alive. A local criminal boss Zorro (or Sorro – the dubbing is unclear) realises gold is at stake and wants Oaks alive for interrogation. Digging the bullets out of him (the 'doctor' remarks 'you won't feel a thing' to the groaning and screaming of Oaks), 'honest citizens' tear him apart when they realise these are gold bullets.

Structured by the 'factional' pattern, with the hero moving between them, we have three 'groups' in the town. The barman Tembler, initially in alliance with the Alderman, but who later split over the gold, creating the 'faction' of Alderman and his mad wife, and finally Zorro, with his black clad and often open-shirted gang, which, Questi points out somewhat redundantly, as *fascisti*. The stranger stays with each of these groupings in the course of the film, moving from Tembler's saloon to Zorro's hacienda, then to Alderman's domestic gothic. In each case these surrogate families are constructed through an hysterical and excessive sexuality: at Tembler's, his son Evan's violent rejection/desire for his father's mistress, Flory, is expressed by his slashing her clothes; at Zorro's, a now kidnapped Evan, being used to extract the gold from Tembler, is sexually-abused, off screen, by Zorro's gang, who have been taught by Zorro to 'enjoy good things'. Evan commits suicide in the morning and, in one of the more sinister remarks in a remarkably sinister film, Zorro says 'He didn't want to be a man... a man who can take on responsibilities, a man who does what he must and accepts it.' Finally, at the Alderman's house the stranger is seduced by the Alderman's deranged wife who, in full Bertha Mason mode, will later burn the house down.

These 'sexual' exchanges are mirrored in the film's use of the stolen gold as the 'object' that inscribes a lack and excess, equivalent to the structuralist *mana*, the dummy hand, Othello's handkerchief, Poe's purloined letter; the empty object that 'circulates' in the structure, and everywhere brings death and passes through death and corpses. Seized in the massacre of the soldiers guarding the shipment, then the execution of the 'disposable' members of the gang, the smelting of the gold bullets from the share interned with the stranger, the 'liberation' of the rest of the shipment through the killing of the bandits, the literal extraction of the gold torn from the still-living flesh of Oaks, the gold which then leads to Evan's sexual abuse and suicide, which is stashed in his coffin, and finally the half share that melts in the fire set by Alderman's mad wife and encases him as a living gold corpse.

The gold has the function of motivational value but, if not quite converted into the Freudian equivalent of excrement, has the levelling, if not nihilist, function of equivalence through death and the 'abusability' of bodies. To use one of Marx's favourite quotes from Shakespeare's *Timon of Athens*:

Gold? Yellow, glittering, precious gold? No, gods,
I am no idle votarist: roots, you clear heavens!
Thus much of this will make black white; foul, fair;
Wrong, right; base, noble; old, young; coward,
valiant.

The 'common whore of mankind' is, precisely, the 'quilting point' (*le point de capiton*) of sexual and social violence, to use the Lacanian term.[23] Gold functions in *Django Kill!* as the 'floating signifier' *par excellence*, it is the term that unifies the ideological field and texture of the film's universe. At the same time, within that universe, we see demonstrated the excess violence implicit in this ideological structure that is usually concealed by the seeming 'neutrality' of money as 'general equivalent'. In Marx's terms gold is rendered as the '*visible God*', but the 'alienated *capacity* of *mankind*', in Marx's words, has no possibility of return or recovery.[24]

UNBROKEN INWARD REBELLION

What we have in these works is the displacement of the epic towards Badiou's inscription of an 'epic nihilism' that is inflected by the passion for the real. That 'passion' is not simply the revolutionary passion, but rather the 'passion' of the everyday brutality and enmity of capitalism. In Engels' memorable characterisation, from the *Condition of the Working Class in England*:

When one individual inflicts bodily injury upon another, such injury that death results, we call the deed manslaughter; when the assailant knew in advance that the injury would be fatal, we call his deed murder. But when society places hundreds of proletarians in such a position that they inevitably meet a too early and an unnatural death, one which is quite as much a death by violence as that by the sword or bullet; when it deprives thousands of the necessaries of life, places them under conditions in which they cannot live – forces them, through the strong arm of the law, to remain in such conditions until that death ensues which is the inevitable consequence – knows that these thousands of victims must perish, and yet permits these conditions to remain, its deed is murder just as surely as the deed of the single individual; disguised, malicious murder, murder against which none can defend himself, which does not seem what it is, because no man sees the murderer, because the death of the victim seems a natural one, since the offence is more one of omission than of commission. But murder it remains.[25]

Still, Sergio Corbucci's *The Great Silence*, 1968

The Spaghetti Western, in its political guise, gives *form* to this violence as literal murder – deriving from the explicit violence of original accumulation a figuration of inexplicit everyday violence.

This experience was raw in an Italy that had witnessed large scale internal migration from the rural South to the newly industrialising North during the 1950s and 1960s. The influx of young male workers, no doubt the viewers Franco Nero had in mind, experienced both a 'late' form of 'primitive' or better, 'original accumulation', and the immersion in the new inertial world of factory labour. The Spaghetti Western, probably inadvertently, mediates this experience that binds together these experiences - displacement, the rural, inertial labour, and the precarious violence that composes the 'rule of (capitalist) law'.

The excess of the Spaghetti Western's violence reveals the violence encrypted in labour: in the subsumption of living labour, the pumping out of value, and the replacement of living labour with dead labour. This 'epic' takes a tragic form; Marx remarks in the *Economic and Philosophical Manuscripts*: 'Wages are determined by the fierce struggle between capitalist and worker. The capitalist inevitably wins.'[26] The Spaghetti Western is the film of defiance in the face of an awareness of the experience of defeat unfolding through militancy and the acceleration of armed struggle.[27] This is a radicalised proletarian sensibility that is not simply a joyous celebration of violence against the bosses, though it is that, but also awareness of the logic of repression, and resistance to the epic tone of prophesying or fantasising victory, and denying defeat, that took hold in certain factions of the movement, armed and otherwise, of the 1970s.

This epic nihilism, given a more elegiac tone in Peckinpah's work, now seems to figure the crisis of labour, a long drawn out defeat, the de-energising of nihilism into the superfluity of labour. In fact, we might revise or question the projective fantasies that could attach to such a sensibility, and see instead something more austere in that excess, a registration of historical defeat in advance that depends on the incorporation of such defeats at the bodily level.

Engels recognised that the violence of the capitalist class resulted in a counter violence:

> There is, therefore, no cause for surprise if the workers, treated as brutes, actually become such; or if they can maintain their consciousness of manhood only by cherishing the most glowing hatred, the most unbroken inward rebellion against the bourgeoisie in power.[28]

His prophesy was that communism would mitigate and civilise this violence, providing it with its dialectic. The communist aim to do away with class antagonisms displaced it from embracing a bloody war of classes: 'In proportion, as the proletariat absorbs socialistic and communistic elements, will the revolution diminish in bloodshed, revenge, and savagery.'[29]

The Spaghetti Western, in the instances I've traced, does not seem so sanguine about this dialectic, and in fact aligns the experience of hatred and nihilism in the experience of defeat that is everyday experience. Lacking faith in the victory of proletarian violence over the technological and politically inflated violence of the capitalist state and capitalist economy it resonates in registering an antagonism, but is less hopeful that the solution to the riddle of history can be achieved.

Benjamin Noys <b.noys@chi.ac.uk> is a theorist living in Bognor Regis. His most recent book is *The Persistence of the Negative: A Critique of Contemporary Continental Theory*, Edinburgh: Edinburgh University Press, 2010. His blog is http://leniency.blogspot.com

FOOTNOTES

1   Gilberto Perez, 'House of Miscegenation', review of *Hollywood Westerns and American Myth*, by Robert Pippin, *London Review of Books*, 32 no.22, 2010, pp.23-26, http://www.lrb.co.uk/v32/n22/gilberto-perez/house-of-miscegenation.

2   Gilberto Perez, *The Material Ghost: Films and their Medium*, Baltimore and London: Johns Hopkins University Press, 1998, p.241.

3   Ibid., p.247.

4   Ibid., p.234.

5   Christopher Frayling, *Spaghetti Westerns: Cowboys and Europeans from Karl May to Sergio Leone*, London and New York: I.B. Tauris, 1998, p.xix.

6   Philip French, *Westerns*, London: Secker & Warburg/The British Film Institute, 1977, p.9.

7   In Frayling, op. cit., p.xi.

8   Anon., 'Let's Do Justice to Our Comrade P.38', in Italy: *Autonomia, Post-Political Politics*, Sylvere Lotringer and Christian Marazzi (Eds.), *Semiotext(e)* III.3, 1980, pp.120-121.

9   Johanna Isaacson, 'You Just Tarried with the Wrong Mexican: Machete and the Aesthetic Politics of Negation', Lana Turner Journal Blog, 2010, http://www.lanaturnerjournal.com/online/49-film/135-you-just-tarried-with-the-wrong-mexican-machete-and-the-aesthetic-politics-of-negation

10 Ibid.

11 Gail Day, *Dialectical Passions: Negation in Postwar Art Theory*, New York: Columbia University Press, 2011, p.3.

12 Frayling, op. cit., p.x.

13 French, *Westerns*, p.43.

14 Frayling, *Spaghetti Westerns*, p.xxi-xxii.

15 French, op. cit., p.43.

16 Alain Badiou, *The Century*, trans., with commentary and notes, Alberto Toscano, Cambridge, UK, and Malden, MA: Polity, 2007, p.85.

17 The 'practico-inert' is a term coined by Jean-Paul Sartre in *Critique of Dialectical Reason* (1960), defined as a field of activity, which, despite being the outcome of a successful struggle by some group, has ceased to be responsive to that group's needs. Bureaucracy is the classic example of a 'practico-inert'. From http://www.marxists.org

18 Maurice Blanchot, "Factory-Excess', or Infinity in Pieces', in *Political Writings, 1953-1993*, trans. and intro. Zakir Paul, foreword Kevin Hart, New York: Fordham University Press, 2010, pp.131-132.

19 Ibid., p.131.

20 Ibid., p.132.

21 Maurice Blanchot, *Lautréamont and Sade*, trans. Stuart Kendall and Michele Kendall, Stanford, CA.: Stanford University Press, 2004, p.68.

22 In Alex Cox, *10,000 Ways to Die*, Harpenden: Kamera Books, 2009, p.143.

23 Slavoj Žižek, *The Sublime Object of Ideology*, London: Verso, 1989, p.87.

24 Karl Marx, *Early Writings*, intro. Lucio Colletti, trans. Rodney Livingstone and Gregor Benton, Harmondsworth: Penguin, 1975, p.377.

25 Friedrich Engels, *Condition of the Working Class in England*, 1844, Chp.7, Marxists Internet Archive, http://www.marxists.org/archive/marx/works/1845/condition-working-class/ch07.htm.

26 Karl Marx, op. cit., p.282.

27 I owe this point to Giovanni Tiso.

28 Engels, op. cit., Chp.7.

29 Engels, op. cit., Chp.13.

FTH:
THE SAVAGE
AND BEYOND

HOWARD SLATER *grasps the August riots as the appearance of an 'unrecognisable demos' which challenges the very ability of capitalist democracy to include or contain the language and acts of its subjects*

1) One thing can be said with a modicum of certainty: the recent riots of August 2011 were political. What can be meant by political in this instance? Well, maybe something as straightforward as taking action in the street, getting beyond the idea of a 'neutrality of living'.[1] It's a form of such neutrality that informs those accusations that have it that the riots were 'apolitical'. These accusations more or less come from a political state (and those professionally invested in it) that proffers an idea of politics as the maintenance of a 'neutrality of living', as the embodiment of rational common sense, as the legislative thrust of a protective equilibrium. If, as many of the riots' detractors maintain, these actions that overspilled the boundaries of 'civil society' were not political then we have further reason to surmise that politics for such as these is a technical managerial affair: the management of libidinal and economic energies into a steady state; the making legible of all action into recognisable 'civil' forms.

2) The next step, then, was to brand the rioters as criminals and to see in the rioting a mass outbreak of criminal opportunism. Such labelling brings the 'overspill' into an understandable civil remit and makes exemplary retaliation possible. Anything else is unconscionable for those who sit comfortably upon us. So, to link the ongoing austerity cuts to the riots as some liberal politicians did, is

seen as an outrage, as a breach to the morally consensual re-establishment of 'civil peace'. This civil peace is however, informed by capitalism's 'naturalisation' within the state as it creeps beyond a simple integration – that could be echoed by a discipline going by the name of 'political economy' – towards a takeover of the state political realm – that could ring out in such a phrase as 'corporate governance'. It is maybe possible, then, to consider the increasing role of the state as an indicator of a corporate takeover – ongoing welfare cuts, bank bailouts, effective corporate lobbying, etc. – that makes politicians 'instinct' with capital and dictated to, in the moments immediately preceding the riots, by the fear of 'sovereign debt'. Thus they are dictated to by a form of corporate para-political power that wields abstract measurements of a state's wealth and economic standing over and above even these states' belief in a mythic 'democracy'.

3) The moral outrage of the governing political administrators at this disturbance of 'civil peace' is, then, illustrative of a kind of 'political theology'. Their god is now capital and those blasphemers against value, property and the entrepreneurial form are sent off to a purgatory of incarceration and classification. In this theocracy it makes perfect sense to jail a youngster for nicking a bottle of water because the greater crime, beyond that of thieving the object itself, is the blatant disregard it shows for surplus value and the exchange value embodied in money. Likewise, robbing for the sake of it not only shows up the worthlessness of these commodities – a disrespect for the commodity as much as an indicator of greed – but it is a kind of people's auto-reduction to

the 'naturalised' criminality of 'markup', 'profit margins' and welfare 'bribes'. Of course, part of the moral outrage stems from the extent and ubiquity of the looting and this outrage must, by self-preservatory necessity, be blind to the 'oversignification' of the riots that, in their very chaos, appear as a kind of 'deforming' mandate that, weirdly enough, complements the politicians presiding over a foreclosure of a by now financialised politics.

4) This 'oversignifcation' is polysemic noise and it is enigmatic to the turnkeys of capital who, seeing the 'neutrality' of their governmental and ideological forms retaliated against, are set adrift before what Miguel Abensour might call a 'political moment': 'the moment most liable to gain an excessive meaning, to go beyond the meaning proper to it'.[2] The meaning deemed 'proper to it', as usual, is criminality, but the excess of this meaning should take in the relationship between crime and poverty, between abandonment and rage, between hope and despair (all inadmissible in a court of law). That the riots are a political moment does not mean that those who participated in them face the political state as an homogeneity; they are not consciously proletarian, though a majority are working class (or even part of the 'surplus population'); they are not all gang members or linked in varying degrees to a more organised crime set up; they have not all been stopped and searched; not all been in prison, etc.[3] The move to deem all who participated as 'criminals' - as if to suggest all those who carry convictions are the same 'type' - is just as much about the refusal to see the politics in hope and despair, in abandonment and rage as it is a refusal to see the politics in poverty. Whilst this latter has a long tradition, it is a tradition that, on the whole, has, like the political state, not taken in the affective

dimension which, were it to do, may have led to a less meaning driven capture of the riots; to a 'thinking emancipation otherwise'; towards a re-forming of 'political links' as relational.[4]

5) The outrage of the political class meets the enragement of those subject to austerity, and whilst they are not an homogenous mass, could it be said that, speaking a 'language of acts' (Pasolini), those out on the streets formed an 'horizon' for the political state as well as for us politcos; and became a form of 'unrecognisable demos'.[5] No longer an 'idea as subject' (a definable gender, class or race) these crowds, often called upon to 'participate', return the loaded inveiglement to speak with a language (as often a blasé beat as a gestural scream) that

---

politicians 'instinct' with capital

is untranslatable into the language norms that would seek to bestill it as 'criminal', as 'proletarian', as 'underclass', as 'materialistic'. Not fighting for a cause, but fighting against causes, against a dimly perceivable - but all the same felt - overdetermination of their lives by, well, in the immediate past, austerity measures caused by the bank bailout. Not fighting as a 'body' but fighting for the body, fighting the pressures felt by bodies in the form of abandonment, hunger, desire, aggression, alienation and stoic hopelessness. In Tottenham, that the family and friends of a police murder victim were ignored after requesting to be heard out is an indication of a callousness that comes along with capitalist social relations: the 'correct channels', the strict form of even a verbal exchange, could not be 'exceeded'. So these non-relations began to be

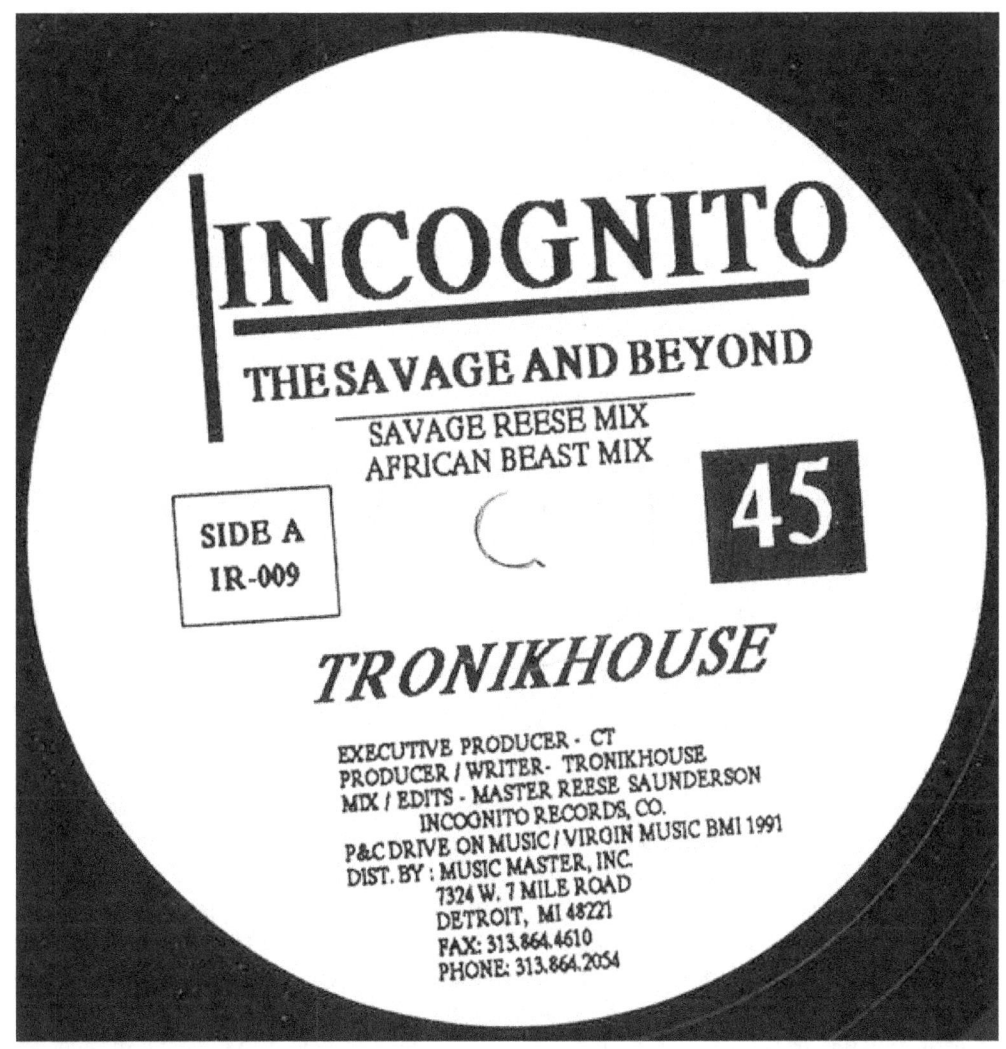

Tronikhouse, 'The Savage and Beyond', 12" vinyl, Incognito Records, 1991

'exceeded' (de-linked) by an at times vicious secession from those very channels; a secession away from the 'policed' language of politics towards appeasing the demands of instinct for which there is maybe no language except the 'language of acts'.

6) Yet instinct, such as rage, is not apolitical. The sexual instinct, the appeasement of which is often negotiatory, a form of communication, is political to the degree that this 'negotiation' of drives and their timing, their relationality, is political. Rage, present in muted form as the aggressive component of the sexual instinct, is a similarly political moment by means of its

---

high streets became the site of an anxiety informed 'language of acts'

---

modulation of transgression and negotiation. So, like the sexual instinct, rage doesn't just come from anywhere (again the criminality tag helps the forces of 'civil peace' to occlude the affective dimension), but its causes are a manifold layering of experiences through which a person comes to feel affronted, neglected and unwitnessed (not negotiated with). When these forms of emotional deprivation (lack of care) meet a situation of poverty and the pressures of material survival through which living horizons and future possibilities are extremely foreshortened, then, in some circumstances, when it is felt there's nothing to lose and nothing to live for, rage can stalk the social psyche. So rage, it could be said, comes to be expressive of a lack of hope, a lack of hope that cannot be countenanced or communicated because, until the circumstances provoking its enragement are met, this rage exists as immanent.

This may well go some way toward getting a handle on some of the acts of 'concise violence' witnessed in the riots – the burning of inhabited buildings, the hit and run, the assault and subsequent death of a pensioner.[6] Even the more embittered revolutionaries would find it hard to condone such 'savage acts', but it is maybe that writers like Pasolini and Genet, who embrace, neigh love, the 'savage' and the wilfully abject in its human form, perhaps it is such writers as these who seek and accept something else in this rage. Both Genet and Pasolini often get impatient with knowing, as many politicos know, that rage can arise at the sight and feel of exploitation, can come from a conscious, felt sense of alienation from the economic and political system, can come from the actions of the police. They sense, too, that both the enraged and the outraged can experience a 'blackout', a short circuit, as impulses takeover. Perhaps this impulsiveness is what comes over when we see footage of folk tearing at the shutters of inconspicuous shops and, as one eye witness described, scrambling in a heap, fighting over a spilled tray of looted jewels. And yet, even in these moments Pasolini, for one, would spare the rioters an empathic hearing and spare us too the trap of our ideolectual urge: they 'appear not only without any logical goal but without even the shadow of an idea; merely expressing with all its strength the general disquiet and restlessness – the anxiety, in fact'.[7]

7) Impulsive? How far does anxiety inform the impulsive? Even so, what can we possibly expect? For Bernard Stiegler, coupling the 'structurally short term' effects of fictitious capital to the lust of the drive for appeasement of needs (instant gratification), capitalism has become drive based: 'novelty is valorised at the expense of durability, and this organisation of detachment

(unfaithfulness/infidelity) contributes [...] to the spread of drive-based behaviours'.[8] Whilst this could possibly gloss some of the more gratuitous looting in the August streets, it also sets this looting against a backdrop of speculative greed and 'instant returns' on investments of the glorified criminals in the banking sector. The very cuts whose blade could have been close to a riotous skin are themselves short termist and the deregulation, in a wicked inverse, was extended to the deregulation, the temporary de-forming, of the law of the land. So, was it that high streets became the site of an anxiety informed 'language of acts'? Did they become the scene for a dissociated revenge of dissociated consumers? Was the libidinal energy so sought after by the window displays returned to them as a 'quick fuck' minus the time of desire? Was it a political indicator that, for some bordering on many, there is no 'neutrality of living' when life itself is no longer guaranteed and relationships are full of hard to express anxiety? Was it that the broadcast effacement of such neutrality has to be called 'greed' in order to ground it in capitalist culture, but, at the same time, remove the desperation of poverty from purview and, furthermore, discredit any claim it may have to inhabit the political?

8) Just as capital is intermittently faithful to those who can pay, so, too, the semi-detached state is not faithful to all its 'subjects'. The resultant evacuation of the political by those learning about their own abandonment is more like an enforced withdrawal. They are barricaded out by a wall of 'political formalism', procedure and financialised jargon that excludes them. Pasolini again:

> The communicativeness of the world of applied science [i.e. politics], of industrial eternity, presents

itself instead as strictly practical. And therefore monstrous. No word will have a sense that is not functional [...] the autonomous expression of a 'gratuitous' sentiment will be inconceivable.[9]

And so, maybe it could be said that many of those who were out on the street in an active anti-form way in August are amongst those who are neither admitted nor desire to be admitted into the political realm but who, speaking a 'language of acts' respond to the 'monstrosity' with the enaction of an 'unrecognisable demos'. Such a demos, as Abensour would maintain in his musings about what the young Marx meant by 'real democracy', is this fissure through which the political state is 'reduced' from its position as '*the* moment of the political', the dominant instant of the political that keeps the 'other realms' of life subordinate and silent; deems them apolitical.[10] The riots, then, with their 'language of acts', their amassed expression of social insecurity, posed a challenge to the authority of the political state in its mission to maintain homogeneity and police the uncontainable overspill. They posed a challenge, too, to those for whom a certain social opacity and indeterminacy are similarly anathema.

Such a temporary loss of civil foundations and the move into the streets of an indeterminate body of people displaying an acute restlessness, an at times 'kamikaze' flouting of the law and a need to be heard in their painful grievance is reminiscent of Claude Lefort's 'savage democracy'. That which, in the guise of the rageful 'raw being' of the sans-culotte of the French Revolution, marked a founding moment of the modern western political state.[11] Moreover, as Abensour informs us, the 'right to insurrection' was deemed to be a 'democratic right' back in those days and, he adds, up to the Paris Commune. That this history, and indeed the events of the Arab Spring, are kept at an ineffectual distance from most western political states, is indicative of the reification of 'democracy' that can admit no further instituting 'discrepencies'. The dream of the rational state as the 'organising form that passes for the whole' now has capital as this organising form and as a result politics, in the form it is now practised (affectless and technocratic), seems more and more to be revealed as the language of a 'pseudo totality'.[12] The 'neutrality of living' becomes asphyxiated in its own forms of empiricised communication and the 'non-totalisable' human becomes thwarted and 'renditioned'.

9) As it could be heard from Tahrir Square, so it is heard from the London prisons that 'our voice has been heard ... we are not animals.' That voice may be indiscernible to some, to others as a 'criminal' voice, it should not be heard and it should be stripped of its 'rights'. But it is a voice that, held in reflux by a structurally informed muteness, may well be speaking the 'language of acts' and seeking to be expressive of affect rather than being 'educated' to speak that language that Jean Genet complained about in the 1940s: a 'language of words' that are 'weighed down with precise ideas'.[13] This weight of the determinate, this exchange value of expression whereby affect (a proto-meaning in itself) should be forfeited for precision with the resultant reward of entry into 'civil society' and its politics, is in itself a bottling up of the 'overspill' of affect and a surplus humanness that engenders the 'savage' as a moment of the 'species-being'. Pasolini and Genet, with their 'telling inarticulacy', well understood this 'language of acts' as a somatic and poetic embodiment, a flouting of the dictatorship of the 'formal universality' of state sanctioned

modes of language. Genet writing of his time spent with the Black Panthers in the early '70s says: 'the force of what was called Panther rhetoric or word-mongering resided not in elegant discourse but in strength of affirmation (or denial), in anger of tone and timbre'.[14] Such 'word-mongering', ringfenced as 'aggressive', and ignored as 'rageful', may well be, in Abensour's view, an instance of an imprecise 'savage democracy'; a political moment in which we can playfully create one another and in which passion as a relational link is once more given a space in contradistinction to a reified democracy. This latter cannot countenance that its 'subjects' are filled with the discrepencies, contradictions and discords of an uncontainable species-life that, in its 'telling inarticulacy', seemed, in August, to bring into view an experience of the wobbling of those very institutional foundations that are charged with maintaining 'civic peace'.

10) Despite their 'criminal' tag, those taking part in the riots formed an opaque body that academic research will now be charged with making transparent. Is this opaqueness, infuriating to the political state, not the very auto-conflictual opaqueness of species life? How often can we be said to be transparent to ourselves? Moreover, is it not an opaqueness through which the enigma of the self and the enigma of the social can no longer be solved accept as a 'language of acts' that entails the abandonment of forms of 'ideality' that have blinded us to the (albeit risky) 'instituting' power of species-being? As Abensour adds in his musing on 'savage democracy': 'Every social manifestation is in the same movement a threat of dissolution, an exposure to division and to the loss of self, as if every manifestation were inhabited [...] by the threat of its own dissolution'.[15] The riots may

well have brought into purview this threat of dissolution of the political state and its 'precise words', they exposed us to division but in so doing they exposed us to divided selves, to a 'self-discrepency' between, say, condoning the most savage acts whilst feeling an optimistic excitement at the breach they formed in the 'neutrality of living'. The riots could be said to highlight such schizo states; states in which the contradictions of living become 'felt contradictions' that not so much bypass thought as bring it into relation with instinct. So, maybe the affective experience of the 'savage' that the riots permitted could get us beyond a reified democracy whose rationalist 'pseudo totality' demonises our 'savage' selves and thus, in line with the myth of productive progress, removes a key facet of our indeterminate species-life and makes us, by means of such devices as guilt, ready to be produced as the financialised subjects required of the political state.[16]

11) This notion of the 'production of the subject' may well figure the riots as a form of 'human strike'. Not only in the suspension of the 'human' in favour of the 'savage', but as a retaliatory strike against the very apparatus of the production of the subject through such institutional *dispositifs* as the education system. For Pasolini, who consistently and painfully spoke of the genocide of the working class, the political state had become 'the new production (production of human beings)'.[17] Maybe this could take on an added resonance in that the increasingly noted failure of capital to reproduce the working class leads to a necessity for the subject to be produced elsewhere than the site of wage labour. This hiatus, this dissolution of the working subject and the incursion of state-led control into the 'bodies' of its subjects, marks a further opaque void for the

the semi-detached state is not faithful to all its 'subjects'

political state as well as the traditional left. The regulatory mechanism of wage labour is absent for many, perhaps more so for those who were out on the streets in August. Just as the looting, then, flouts the law of the wage relation, so too does the latter's absence remove a main social identifying pole. That for some this is to be welcomed (*ne travaillez jamais*), for others it may be an instance of the 'loss of the self' when the 'self', under the value-form of capital, is encouraged to identify with the various roles that wage labour allots. The 'meaning of life' is bound up with work but its absence throws us back on a kind of 'savage' survival and an equally savage interrogation of what it is to be human (a social individual) without the capital imposed definition of life as a life of wage labour.[18]

As Abensour is quick to point out, the 'savage', as Lefort uses it, is not the return to a state of nature, but a baseline in the forecoming of *'species existence, the advent of human existence'*, a socialised nature that must admit of 'other realms', those other disavowed areas of species activity that should not be compressed out of existent expression by the political pseudo-totality of a reified and financialised democracy.[19] The inadmission of the 'savage' to the demos is, of course, one sided. The savagery of financial capitalism is, perhaps, due to its level of abstraction and its ultra-sublimated mechanisms, admitted poll position (the mechanisms of both are a snug fit). But the 'beyond' of such savagery is, as Pasolini mourns, nothing less than a genocide, nothing less than the sacrifice of 'raw being'. An element of this 'raw being' was sacrificed to the law courts in the weeks after the riots. It had no 'idea' to defend itself with, only 'gratuitous sentiment' and the 'language of acts' that disqualified it from the polis and made its 'unrecognisable demos' truly opaque. Its sudden speed of arousal left

us aghast and, at times, thankfully speechless. Theory, always playing second fiddle to praxis and affect, comes in to temporarily save us: 'The "ceremony" of the political should be converted to the species-life of the real and total being of the demos [...] that in its people-being belongs at once to the political principle and the sensualist principle'.[20] There is, then, some urgency in at least revealing the pseudo-totality of politics to its affectless practitioners (who, as Pasolini maintains in his intervention at the Radical Party Congress of 1975, 'live their rhetoric with a total absence of any self-criticism').[21] Such a pseudo-totality, the happy enterprise of unreflexive selves in an unquestioning relation to the 'all' of their knowledge, creates a politics without the sensual and savage component of species life.[22] It is a stunted totality passing for an *ad hominen* practice; the remainder that could well overspill the totality with its 'savage democracy' goes by unnoticed and so too does the 'can-be of the self-contradicting'.[23] Pasolini and Genet had already begun to speak of such an impossible being as a becoming. Beyond the love offered by them to the 'people-being', their love of the remainder and the remaindered made them hopeful that a 'politics of anxiety', a politics premised upon an acceptance of both the savagery of species-life and the radical surprise of indeterminacy, would go some step towards blasting up this lifeless ceremony that passes as politics.

Howard Slater <howard.slater@googlemail.com> is a volunteer play therapist and sometime writer. His book, *Anomie/Bonhomie & Other Writings*, will be published by Mute Books in January 2012

DISCOGRAPHY

Tronikhouse: 'The Savage and Beyond', Incognito
Records (1991)
http://www.youtube.com/watch?v=xkCKjEUhiAM&feature
=related

The Pop Group: 'Thief of Fire', Radar Records (1979)
http://www.youtube.com/watch?v=p6Vvlk7UbEo

Newham Generals: 'Things That I Do', Dirtee Stank
Recordings (2009)
http://www.youtube.com/watch?v=6AUkh-UNhFU

FOOTNOTES

1   Miguel Abensour, *Democracy Against The State*,
    Cambridge: Polity, 2011, p.34. Much of what
    follows is informed by this book in which Abensour
    conducts an exploratory reading of Marx's *Critique
    of Hegel's Philosophy of Right*.
2   Ibid, p.57.
3   For a take on Marx's concept of 'surplus population'
    see 'Misery and Debt – On the Logic and History of
    Surplus Populations and Surplus Capital', *Endnotes*
    II, April 2010.
4   For 'thinking emancipation otherwise' see: Abensour,
    op.cit., p.vii.
5   This 'language of acts' is maybe far distant from
    a rhetorical and more acceptable 'speech act',
    it is an embodiment, perhaps in this instance,
    of a spontaneity (itself conditioned by affective
    layering?) that subtracts from a 'higher' yet
    blunted rhetorical explanatory plane and becomes
    expressive of a suffering (*phônê*) that is too
    'savage' for the *demos* and hence 'unrecognisable'.
6   The phrase 'concise violence' is Genet's, see: Jean
    Genet, *Our Lady of the Flowers*, London: Panther,
    1965, p.248.
7   Pier Paolo Pasolini, *Petrolio*, London: Secker &
    Warburg, 1997, p.436.
8   Bernard Stiegler, *Towards a New Critique of
    Political Economy*, Cambridge: Polity, 2010, p.83.
9   Pier Paolo Pasolini, *Heretical Empiricism*,
    Washington: New Academia Publishing, 2005, p.34.
10 Abensour, op.cit., p.92.
11 Ibid, p.102-124.
12 Ibid, p.60. John Holloway puts it: 'The state,
    by its very existence, says in effect, "I am the
    force of social cohesion, I am the centre of

social determination."' See John Holloway, *Crack
Capitalism*, Pluto Press, 2010, p.133.
13 Jean Genet, op.cit., p.72.
14 Jean Genet, *Prisoner of Love*, New York Review
    Books, 2003, p.56.
15 Abensour, op.cit., p.104.
16 Such musings could be read as 'primitivist' if we
    were to believe that our 'indeterminant species
    life' is not stock-full of social determinations and
    conditioning that it is risky to express. The risk
    emanates from at least two sides: the academic
    left's wariness of the 'irrational' and the political
    state's fear of the unraveling of our capitalist
    conditioning.
17 Pier Paolo Pasolini, *Lutheran Letters*, London:
    Carcanet, 1981, p.109.
18 Becoming a 'social individual' is itself traumatic
    (savaging our self) when subjects are produced/
    taught to identify as individuals and have
    proprietorial conditions of worth. See for instance
    Lacan's formula: 'I is a fortress' cited by Catherine
    Clément, *Syncope: A Philosophy of Rapture*,
    Minneapolis: University of Minnesota Press, 1994,
    p.122.
19 Abensour, op.cit., p.73.
20 Ibid, p.71.
21 Pasolini, op.cit., p.121.
22 In some senses the Consciousness Raising Groups
    of the '70s Women's Liberation Movement were
    forums for just such a critique of the pseudo-
    totality and for an interrogation of the place of (a
    remaindered) gender within capitalism.
23 Ernst Bloch, *The Principle of Hope Vol.1*, MIT Press,
    1995, p.225.

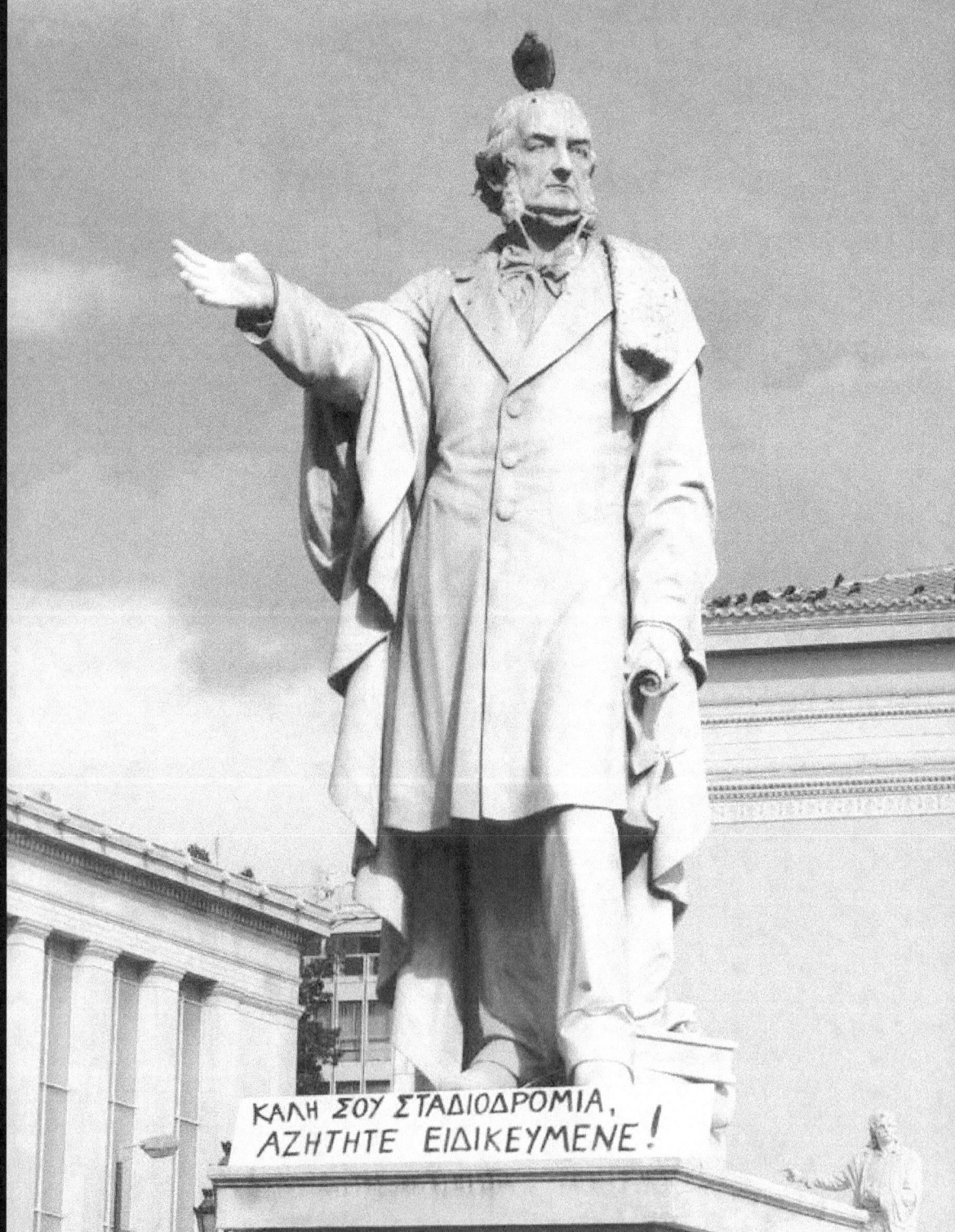

ΚΑΛΗ ΣΟΥ ΣΤΑΔΙΟΔΡΟΜΙΑ,
ΑΖΗΤΗΤΕ ΕΙΔΙΚΕΥΜΕΝΕ!

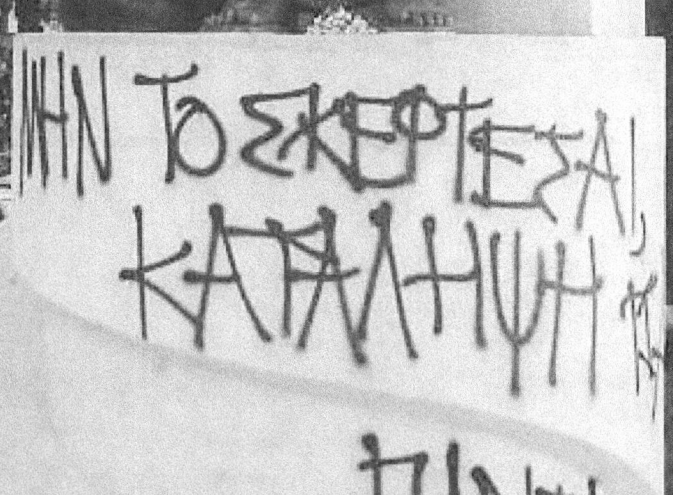

'Don't think about it too hard. Occupation forever', Athens

PAST
CARING

A study day on Motherhood, Servitude and the Delegation of Care at Birkbeck University in May 2011 provoked two writers to examine the politics, pieties and pain clustering around reproductive labour. MADAME TLANK contests the premise that carers must care, however subversively framed, while MIRA MATTAR cuts into the modern Master/Slave dialectic of the Gumtree 'au pair wanted' ad. Finally TLANK assembles careworn voices in a multi-cautionary verse

The family is the basic unit of government
— Michele Bachman, US presidential aspirant

MaMSIE is a fairly new journal (three issues in the past two years) published under the Birkbeck umbrella, attempting to explore and propagate a new discipline of 'studies in the maternal' – MaMSIE stands for 'Mapping Maternal Subjectivities, Identities and Ethics'. This informal study day (participants almost all female and white, plus two screaming babies, n.b., no creche was provided!) attempted to bring together several perspectives on the subject of servitude and the delegation of care.

The atmosphere was very good, with people actually listening to one another and trying to expand on what they heard rather than competitively showing off individual research. It was starkly different from other conferences, which are often male dominated. But even so, the study day didn't quite come together. Although the philosophical, historical and sociological perspectives all focused on the same subject, they dealt in observation rather than analysis and investigation, failing to ask some very basic questions. At its worst it felt like mothers discussing their personal experiences of motherhood.

An exception was Stella Sandford's philosophical enquiry into 'maternal labour', which served as the opening keynote talk. Her attempt to think 'the maternal' within a Marxist framework was interesting, if problematic, for

several reasons, some of which Sandford herself was well aware of and sought to address. She thought that the 'lived contradiction' at the heart of the concept of maternal labour (that I will come to shortly) might help rethink Marx's concept of labour and, more speculatively, that it might be a sign of an inherent non-subsumability of the maternal under capital.

Although she articulated this in a beautiful trajectory of thought and was also well aware of the thin line she was treading, I can't help but ask: why? Why would it be desirable to come up with such a concept of maternal labour and such a conclusion? Why locate the maternal outside of capital and history, so to speak? As though some deep-seated, pre-capitalist residue attached itself to the mother, implying 'traditional' models of maternity? Does it not make it even harder to analyse materially a mother or carer's position vis-a-vis capital, and render it impossible to tear at capital's tentacles from this presumed 'outside'? Does it not delegate to some higher plane the subjective investment in labour which is well known to have been demanded increasingly of workers of all kinds over several decades?

The lived contradiction within the concept of maternal labour according to Sandford is that the mother/carer is both a capitalist subject and a maternal subject. But the maternal clashes only with the crudest caricature of the Marxist understanding of 'labour'.[1] Sandford points to Marxist feminist analyses of domestic labour, but sets the work of child rearing apart from cleaning, cooking etc. due to the emotive investment in, and arguably more rewarding nature of the work. Yet the work of child rearing has the same reproductive agenda as these things: it quite literally replenishes the work force, or reproduces labour power. Childcare is either directly waged, or unwaged and 'paid for' by a wage (be it a man's or the mother's own), or by state benefits (the 'social wage'). The carer's work cannot be separated from the capitalist labour relations that make up her life and measure out her subsistence.

No matter how much 'caring' there is in care work, no matter how much love is invested in the work of bringing up a child, these emotions do not arise in some pure vacuum, free of the imperatives of capital. Furthermore these emotions are channelled and redirected through state intervention in order that their object – the child – may 'mature' into something meeting labour market specifications. The state is more receptive than ever to business lobbying on the form of reproduction and training.[2]

The maternal referred to in this study day and by Sandford is quite a recent concept: the loving, caring mother is a product of, and expectation within, the nuclear bourgeois family. Throughout most of modernity, mothers were by no means automatically expected to care for their children, no matter what class they belonged to. As Shulamith Firestone showed in 'Down with Childhood', in Europe up until the 19th century (and in some areas even the 20th) a poor mother's children were given to wet nurses almost right after birth; they played in packs in the village squares and started working life as soon as they were physically able, doing work (in factories, shops, on the streets) that today would be called 'child labour', or learning trades as apprentices.[3] Rich mothers would give

birth and then pass their children on to wet nurses and tutors and not have very much to do with them either. The presumption of the caring, protective mother is itself a product of fairly advanced capitalist society. And it goes hand in hand with the concept of the child as an unfinished creature, in need of protection, control and care. The emergence in the 19th century of 'childhood' as a separate stage of life, and children as separate (not yet) human beings, reinforced through the eventual separation of children from the 'adult world' by mass primary schooling, developed into what can be called 'abstract childhood'.

A child possesses (or rather is possessed by) childhood for the sublimely tautological reason that the subjectivity ascribed to her is determined more by the single, shared attribute of being-a-child than by all the particular differences *between herself and other children*.[4]

From the 19th century onwards, the child is treated as an incompletely formed subject, requiring special care and proper guidance towards the eventual attainment of serviceable adult identity, until which time precocious adult (i.e. individual) traits are a matter for discipline or therapy, according to various kindergartens of thought.[5] The requirement to care for and 'shape' the vulnerably deficient or 'innocent' child pushes the mother into the work of mothering, i.e. she is made into a *carer*, with her role and that of the child reinforcing one another.[6]

Subjective deficiency is also ascribed to the disabled and the elderly in particular, but at the same time it is projected *in general* across an ever more provisional adulthood. A decades long drive (supply side Thatcherism, the Third Way, the Big Behavioural Society) to hold individuals 'responsible' for whatever economic, legal or medical position they find themselves in has been shadowed at every turn by the assumption that the same individuals are incapable of making the 'right choices' without psycho-managerial nudging or outright institutional coercion.[7] This presumption is applied with particular ferocity to mothers/carers and their ability or otherwise to care 'appropriately'. The extension of abstract childhood to mothers and other adults whose job it is to provide care is evident in their subjection to continuous training and retraining, surveillance and expectation.

The historical papers presented by Kate Pullinger and Lucy Delap on the study day showed how something similar played itself out in Victorian times: as a rule, servants were subject to infantilisation, their assumed place was on a level with the employers' children.[8] The employer had ultimate control over both groups and directly controlled the servants' care of the children. Economically speaking, children and servants were both completely dependent on the master of the house. Sadly there are few incidents known where servants and children ganged up together against their infantilisation and oppressive lives. Firestone's call for mothers to gang up with their children has likewise gone unheeded.[9]

Rather than arguing, as Sandford does, that maternal labour is distinct from other labour because of the element of caring for the object of this labour, might it not be a stronger feminist argument to say that a care worker may be just as indifferent to the object of her care as a factory worker to the product of their labour? Care work does *not* need to be done with care, and often can't be done with care, even if workers wanted to. With some of the lowest wages in the labour market and a workload to kill a (wo)man, there wouldn't be much time to invest in care. The

assumption that the work needs to be done with care is moralistic and unhelpful at best, but at worst it forms an oppressive imperative: in waged labour, care workers are expected and under pressure to care (and control where applicable) under dismal conditions. This pressure usually comes directly from the (private and/or state) employers or the 'customers'. In

---

Regardless of how much care there is in care work and love there is in child rearing, these emotions do not arise in some pure vacuum, free of the imperatives of capital

unwaged maternal labour, mothers are put under pressure to exert a certain level of care (and control) over their children. If this is not done 'well enough', the child might be taken 'into care'. State management of reproduction is more ambitious than ever.[10]

The institutional care of old people in the UK provides clear empirical evidence against the argument that maternal – or, by Sandford's own extension, more broadly caring labour – might somehow contain a remainder that is not subsumed under capital. Recent events make it painfully obvious that the whole *form* of caring – not just its financialised background, but the way the everyday labour is performed – is directly determined by capital transactions of the most abstractly speculative kind.

Leveraged capital started flowing into the care homes sector around five years ago when 'investors and lenders' thought the value of the businesses could only increase as the population aged.[11] Banks were 'happy' to lend to companies with 'steady income streams from government contracts' (50 percent of care home places are municipally funded) and plenty of real estate; all the more so because the debt could immediately be sold on in securitised form. 'It was a bit like the internet bubble' or 'the economic surge generally', said the CEO of Care UK (Bridgepoint Capital). Meanwhile the private equity owners would sell the underlying real estate at a premium obtained by signing up to automatic annual rent increases, using the proceeds to expand the businesses through more buy-outs and then sell them on for 12 or 14 times *projected* earnings. This model left little reason to invest in labour and facilities while the boom continued, but investment became altogether impossible for second or third generation owners as rents continued to rise automatically after credit dried up in 2008 and state payments were subsequently frozen or cut.

This non-investment is no abstract matter. Long before Southern Cross (bought, inflated and sold on just in time by Blackstone) collapsed and *The Financial Times* noticed the story, *Private Eye* had spent years reporting each case of patients dying dehydrated, stuck to the bed by untreated pressure sores and covered in their own piss and shit. This is not a situation that care workers (paid less than £7 an hour at 'senior' level in the private sector or as much as £8.90 in those homes not yet divested by councils) can improve through a supplement of uncapitalised 'love'. Not when a ratio of seven carers to 89 'high maintenance' patients is normal, or when work schedules, available supplies of food and medicine and all aspects of the physical environment depend on the financial position of over-leveraged 'owners' and the state. One telling common experience of carers and residents in these institutions, according to the FT, is that neither would dare report conditions to the regulator's inspectors:

'they're scared, because they're dependent.'

The private market in care for the elderly is mediated throughout by the state, most obviously as wholesale purchaser of services and source of pension income, but also, particularly in the UK, due to years of legislation cultivating a private pensions industry and asset price inflation in private equity and real estate.[12] And in the end, the state will probably have to bail out 'care homes' just like other branches of finance capital before them. (It's also worth noting that government plans for the future delegation of

---

Subjective deficiency is also ascribed to the disabled and the elderly, but at the same time it is projected in general across an ever more provisional adulthood

---

wider NHS services closely resemble the private care home model, slow and predictable crash notwithstanding.)

The state likewise does much to define the form of maternal labour in the narrow sense, albeit less visibly and with the emphasis more on policing than financial brokerage. Welfare state services supposedly supporting single and some other mothers are inseparable from control and surveillance mechanisms, to be endured in exchange for a meagre cash payment. Education increasingly incorporates employers' *qualitative* demands for future labour power, building the desired psycho-social 'training' into curricula.

To a degree connected to their individual level of state dependency, mothers and other unwaged carers are faced with the imperative to make sure their children become suitable future labour power, (the middle classes, meanwhile,

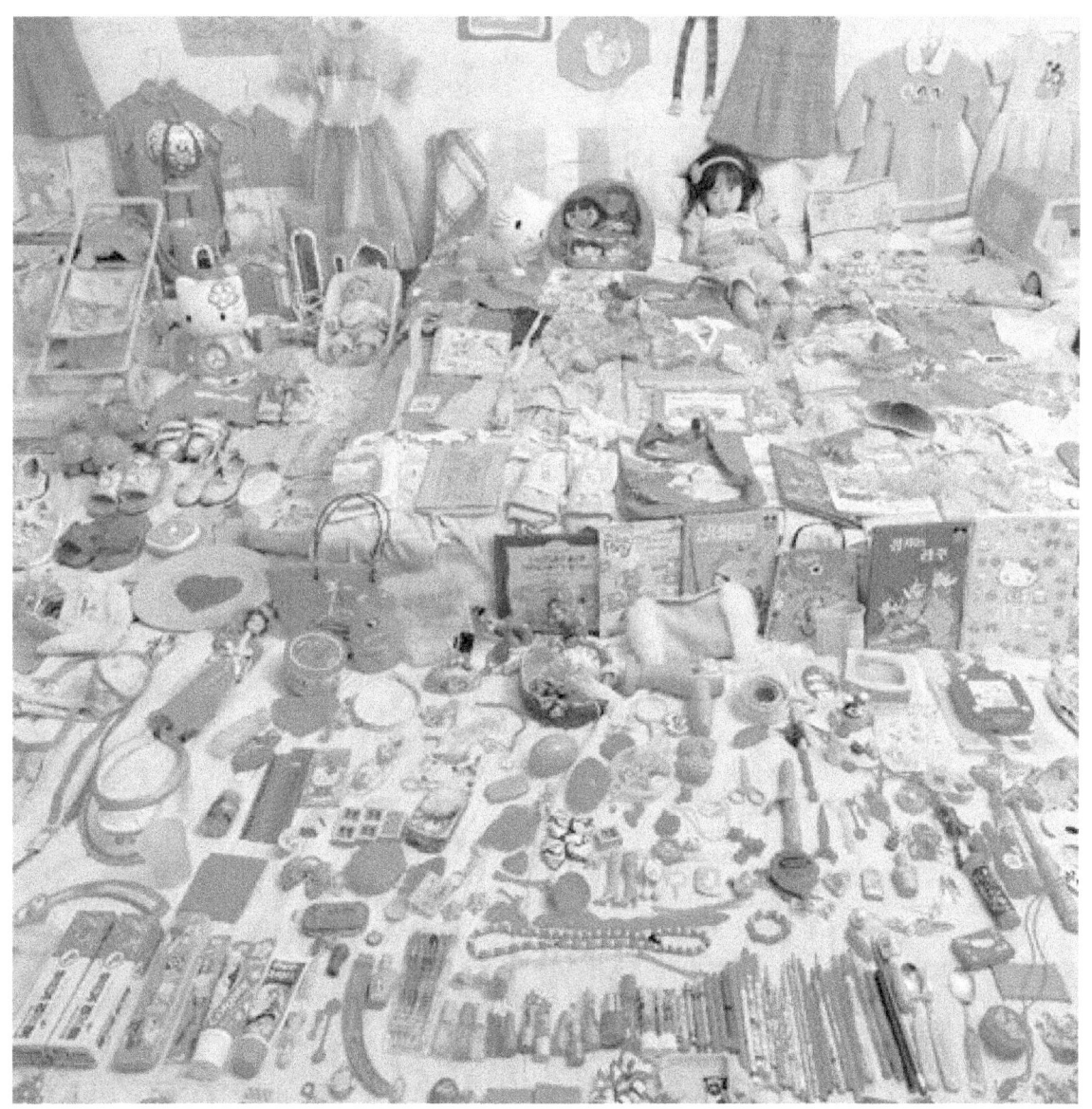

JeongMee Yoon, *The Pink Project - SeoWoo and Her Pink Things*, 2006

internalise these same imperatives and hold themselves responsible for achieving 'desirable outcomes for their children'). At the same time private capital permeates the 'frontline' provision of welfare services, with delegation of medical, benefit and housing functions to PFI-type contractors (paid by 'results!') conditioning the relation between mother and child in manifold ways.[13]

Motherly work means having to make sure your child *adjusts* to reality as constituted by capital. Women who depend on the state for money are more exposed than others to state interference (itself often subcontracted to private capital) to this end. Motherly love can change

---

And in the end, the state will probably have to bail out 'care homes' just like other branches of finance capital before them

---

just as much according to questions of rent (for example) as a commercial care service will. Low levels of state financial support – or low wages part funding unwaged care – are just as much of a hindrance to the subjective investment in care as is an hourly wage of £8. Single mothers (along with disabled recipients of care) are pushed back into competing for work, widening the labour pool at employers' disposal and thereby dragging wages and conditions downwards. This hits mothers on benefits especially hard as it is yet another pretext to control what they do and *how* they care; it also requires them to take part in capitalist labour relations from the opposite side, effectively becoming employers themselves by sending their children either to a nursery or to a paid carer. All these instances show maternal labour to be subsumed, but none

of them stop a mother (or child) from being *in and against* capital.[14]

Sandford's theoretical attempt to show a romantic revolutionary gap in the capital relation, to show the maternal not to be fully subsumed under capital, may arrive at something quite opposed to what she was hoping for.

If the mother or carer's full involvement in capital as a social relation is not acknowledged, then how can we analyse and understand it in order, ultimately, to fight and break it? (Not even to speak of the counter-productive effect of setting the mother/carer apart from other workers.) And the same holds for the state's role in private capital. Reformists' calls to enlarge the role of the state completely ignore the fact that rather than some 'temperate player regulating the wild excesses of capital' the state is a co-player, writing the rules and facilitating the cash flow for a smooth ride on the wild, exciting seas of people's bloody lives.

LOOKING FOR A REAL-LIFE MARY POPPINS!

Dear Au Pair,

I am Andrea and my partner is Christopher. We have two boys, Seb who is seven and Barnaby who is five. They both attend the local primary school which is a ten minute walk from our lovely home in Hampstead, North London. After school they keep busy with numerous activities – football, swimming, gymnastics, music and street-dance. Christopher and I are lawyers and work full-time. When we're not working we love nothing better than to spend time with the family. We are a very warm and welcoming family who have loved having an au pair. Our last au pair was with us for more than a year, she became part of the family within days and will be a lifelong friend.

As I said, our house is lovely and very close to the city centre. There are lots of buses and tubes and we even provide a bicycle. Christopher and I cycle everywhere. Ideally we'd want a fit, sporty au pair who feels confident cycling on London's busy streets. A helmet is a must! So no vanity please ladies! I should say here we're only looking for female applicants.

It is also important that you can take on extra hours in school holidays which we will, of course, pay for, though we do try to take as many of their holidays off as we can which means that you should get plenty of holidays yourself! We never work in August, always take two weeks off at Christmas and a week at Easter – so that's seven weeks already. Your normal working day would be 08:00 to 19:30.

Your duties with the children:
* *wake, feed, dress the children and take them to school (20 minute walk)*
* *collect them from school and activities*
* *help the children with their homework*
* *take them swimming at the weekend*
* *cook healthy, simple meals for the children (we provide recipe books)*

Other duties:
* *clean and tidy the house*
* *do the family's washing and ironing*
* *provide me with an extra pair of hands*
* *feed and walk the dog*

What we can offer you:
* *a large room*
* *a private bathroom (to share with the children, your rooms will be linked via the bathroom)*
* *wi fi access throughout the house (if you don't have your own computer you are welcome to share with us and the children, but I should warn you they do like their computer games!)*
* *We do not provide a television in your room because we want you to feel at home as part of our family and use the TV or watch a DVD in the living room*
* *You can use the telephone anytime you want (if you write down your calls on the pad next to the phone we can easily deduct the cost of your international calls from your weekly wages)*

* *wonderful food*
* *international travel with us on holidays*
* *£120 per week*

You must be:
* *between 20 – 40 years old*
* *fluent in written and spoken English*
* *a confident driver*
* *energetic, sporty and have a sunny disposition (even in London's wet weather!)*
* *flexible (sometimes we come home late from work and need you to be there, we will pay for extra hours)*
* *quiet, neat, clean and showers at least once a day (you'd be surprised!)*
* *an enthusiastic but tidy cook, all our meals are cooked from scratch!*
* *will love our children and help them discover the world*
* *a non-smoker*

Ideally we want someone who is:
* *confident but not domineering*
* *not embarrassed to sing and read with funny voices*
* *warm, patient, supportive, perceptive*
* *creative with their own ideas of things to do with the children that they will all enjoy*
* *a dog lover*
* *at home in museums and galleries. With London's wealth of cultural life we don't want our little rascals to miss out! It will be fun for you too*
* *a modest dresser, with strong family values*
child-free and single as we have found emotional ties at home can be distracting

Yes, we have high standards and yes, such people exist because we have had them in the past! It is one of the most important jobs a person can do. It will be very rewarding for you. We are looking for more than just someone to care for our children, we want you to be part of the family. Please send a CV and covering letter telling us a little about yourself and your background and why you think you'd be right for the job to the following address...

(As you can imagine these positions are very popular so we will only contact you if we want to arrange an interview.)

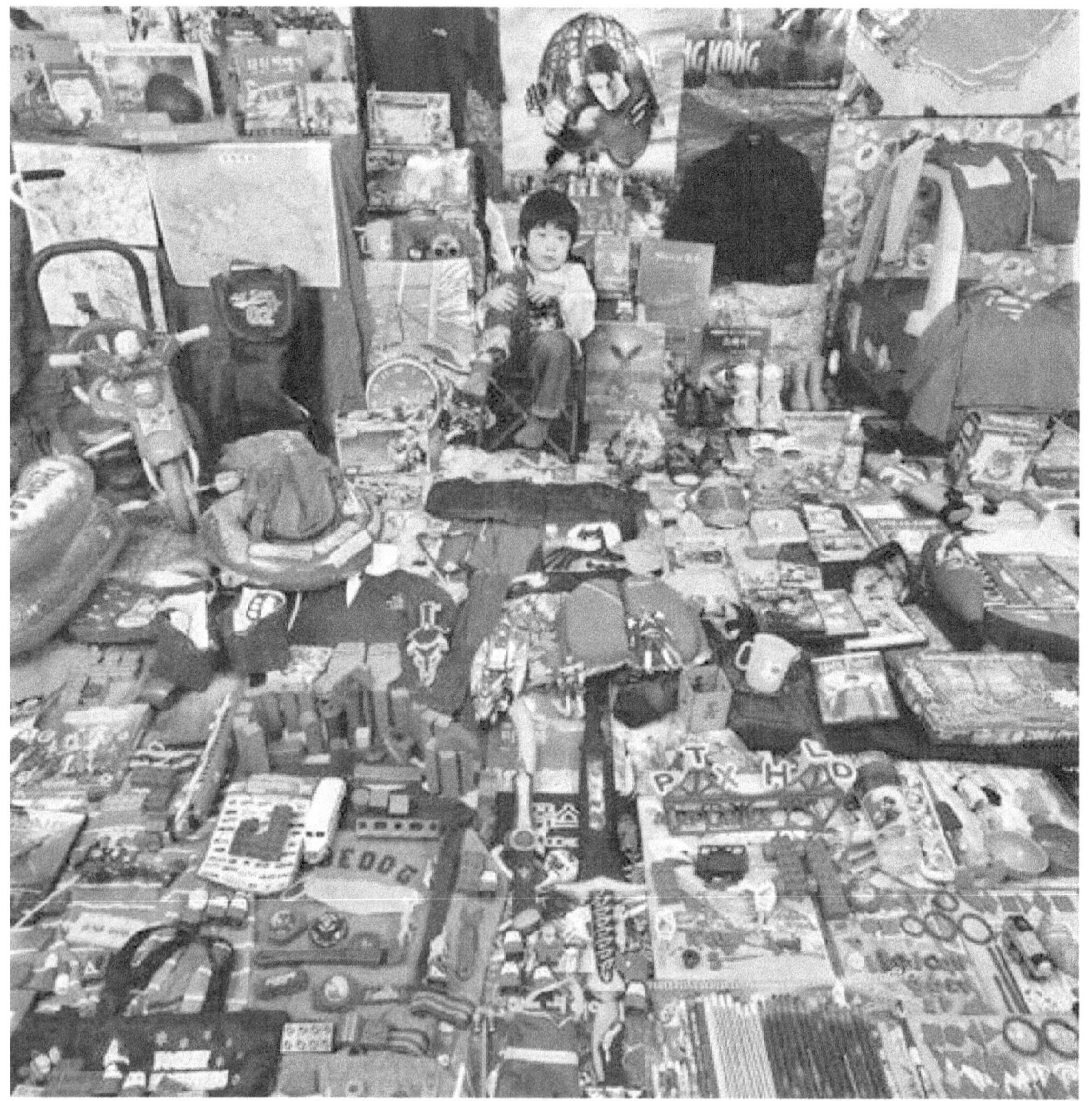

JeongMee Yoon, *The Blue Project – Seunghuyk and His Blue Things*, 2007

Dear Host Family,

My name is Martina and I would like to be your family's au pair. I am from a good neighbourhood in Košice, a big city in The Slovak Republic. My mother is a nurse, my father is an engineer and my sister Lenka is studying to be a doctor. I myself have studied at the Academy of Education and Social Science and have taken A-Levels in Slovakian language and English language. I also speak German. As you can see we are a very hard working family who takes studying very seriously.

I also love to have fun, especially with sports. I love tennis and running and to ride bikes. I look forward to cycling on the exciting streets of London.

But please, do not worry, I am very clean. I shower daily and use deodorant twice a day. I also change my panties and socks every day. I change my bra once a week. I have a different bra for sports. I wash this after every use. I am very sporty and would like to ride a bicycle with a helmet. I have short, easy hair.

I love children and would like one day to have my own – but not soon so please do not be concerned about this. If you are worried let me tell you I have been taking the pill named Microgynon 30 for 12 years with zero problems. I can assure you also that my menstrual cycle will not affect my moods or behaviour, I am very stable and happy. If you are worried still I can have the IUD fitted which the doctor says is more effective than sterilisation. But let me tell you I take strong anti-depressant pills which means I suffer almost no sexual urges whatsoever so I can promise you I will be 100% abstinent during my stay with you (and I do not find the English style of man is for me!). Even still I can take pregnancy tests every six months as many of my friends do in their host families. Just to be sure.

I will love your children very much. I wake up early and like to get out so taking the children and the dog to school sounds great and believe it or not I like the rainy weather too! So this is not a problem. I have a small problem with something called OCD but it allows me to clean very well, so if you leave your home in my hands it will always be spotless. Ironing relaxes me. I love to cook, it is my secret dream to cook in a good restaurant so please allow me to make fun and healthy meals for the lovely children, and even the parents too!

Also because I am a Leo I am a very loyal girl. I will be totally devoted to your children. I will not even look at another child. If another child falls over and damages itself in the park, and blood comes outside, I will not help it. I will quietly monitor the friends the children have to see if any of them are taking advantage of your clever boys! I do not like copy-cats.

I will not be too emotional when it comes to missing my own family. I will be in silent, almost physical pain because of the distance but due to modern technologies like Skype I can keep in touch with them. It is ok for me that you can never meet eyes on Skype. My loneliness will be hidden from the children very well. I am a keen swimmer.

I love to look at art and old things in museums and am so excited to see London and all the beautiful things. Do you know much about Slovakian art? I will be happy to tell you if we meet!

I have heard from my friend of the Ikea style of furniture you favour. I know this type is easy to clean with a dry cloth or rag. I can provide my own rag/s. I look forward to ample storage.

To me, being an au pair will be fun and a challenge and I want to see the world and how other families live. My English is very good but I want it to be perfect, just like your children will be if they have me as an au pair!

I am looking forward to hearing from you, thank you.
From

Martina.

P.S. I see you say the bathroom is shared, this may be a problem for me. I do not want to be un-ladylike but I need to tell you I suffer from IBS so I need an en suite toilet. My movements are painful and I have also a flatulence. This will be made worse by the change in my environment. I hope you understand my need for private toilet facilities. If this is not possible perhaps the children and I can arrange a rota.

'LABOUR' cries child cries labour cries tears
push push!
that's not labour, that's nature, darling
help me, look after me, it hurts!
here pills, must move on, next mom
'Mom' cries child cries it it
must dry tears – this the object of my labour?
maths exercise (you child, must succeed, try once more)
count hours x o pounds x stopped heartbeat
this the value of my labour
'labour labour' I will pay you to perform
on the object of my affective labour by the hour in time
to run
with the school run run flow forward why, you ran?
ah the grandmas count their hours well then you may
earn.
change your standing participate in the accumulation
chaaaain

wait – are you the reason I don't get paid?
but I do – the state throws some goodies at me
and wants some names dates and cleanest of clothes.
good behaviour, watch what parents are up to... outside
the accumulation
chain? for you maybe, well-bred few. not those who bear
and bear and bear
and feed and feed and feed and fuck had enough of
producing your pension
or take it away, hah
this, dear, threads itself through the ages. the workhouse,
the servant,

childcarer, and you (no clue). as long as you don't try to
have them too. must
be on one standing with them helpless little lots. I have
power to pay you or
if I wish not. neither the little ones nor old ones get cash.
you're all of the
same lot, no power, not not. but go ahead fighting
amongst your old selves.
it's called reproduction of the capitalist relaaaations. hah.
there's nothing idyllic in this kind of shit. now I can afford
it I delegate it.
and yous will keep doing the crap. so your little ones grow
big. become like
me and pay the likes of yous for the same old doubletrick.
outside of the sphere you say? all data proves you wrong,
dear. whose
property are children? not yours, they're the state's to
throw into the dirt pit
whenever seen fit
yous, you, your little tiny shrieks. don't you know your
greatgrandma was
an immigrant too. The asylum saw the last of her, past
caring indeed. just
cause you can buy yourself out/ don't be so sure now about
your little
emotive clout.

_____

Mira Mattar ‹miramattar@gmail.com›‹twitter.com/miramattar› is a sometime governess, freelance writer and contributing editor to *Mute* and *3:AM*. She lives in South London and blogs at http://hermouth.blogspot.com/

Madame Tlank continues to stir up the ‹s› in front of care

FOOTNOTES

1   The very fact that Marx finds it necessary to qualify waged/unwaged, productive/unproductive etc. labour as such should be enough to establish that the category of 'labour' in general is not identical to the waged and productive.

2   Impersonate an employer and submit *your* wish list at, http://www.businesslink.gov.uk/bdotg/action/detail?itemId=1086319141&type=ONEOFFPAGE. The Browne Report on higher education is one high profile product of this kind of receptiveness, http://www.telegraph.co.uk/education/8058285/The-Browne-report-in-full.html

3   Shulamith Firestone, 'Down with Childhood' in *The Dialectic of Sex: The Case for a Feminist Revolution*, London: Paladin Press, 1972.

4   Anonymous, *The Bio-Power Digest*, 'Endless Youth', 2004, privately circulated.

5   A prime example of this can be found in the 'Letting Children be Children' report by the Mothers' Union (sic), http://www.themothersunion.org/letting_children_be_children.aspx

6   Cf. the proliferation of child rearing manuals on how to best 'improve' children, especially during the first three years of their lives.

7   For a succinct account of Richard Thaler's 'Nudge' theory and its adoption by the UK government's Behavioural Insight Team, see Michael Fitzpatrick's 'Public health and the obsession with behaviour', *Spiked*, May 2010, http://www.spiked-online.com/index.php/site/article/8785/

8   See, Firestone, op. cit.

9   See Mme Tlank, 'The Battle of all Mothers', *Mute* Vol2 #9, Your Five a Day!, 2008, http://www.metamute.org/en/The-Battle-of-all-Mothers

10 *Financial Times*, special reports, May 31 June 1 2011.

11 For a comprehensive account of this see Rob Ray, 'The 3 Ps – PFI, Private Equity, and Pensions', *Mute* Vol2 #6, Living in a Bubble: Credit, Debt and Crisis, 2007, http://www.metamute.org/en/The-3-Ps

12 See Mme Tlank, op. cit.

13 A rare example of professional jargon actually proving 'fit for purpose', to the extent that it acknowledges the war waged every day in benefit offices.

14 The question of struggle 'in and against' capital is taken seriously in the feminist work of *Big Flame* (1970-1984), now archived at http://bigflameuk.wordpress.com/

RIOT
POLIT-ECON

In this riposte to the moral backlash against the UK's August riots, THE KHALID QURESHI FOUNDATION & CHELSEA IVES YOUTH CENTRE survey the 'immiseration industries' ushering in the emerging regime of work/fare and worklessness

On 17 August, Mayoral buffoon and sometime *nth* grade littérateur, Boris Johnson, announced his official response to the riots. Despite his allergy to accounts of 'social and economic justification', Johnson took the occasion to propose a supplementary £20 million fillip for his £50 million London Wide Regeneration programme. Not only this, Johnson continued, but the Mayoral Office would be disbursing a still more philanthropic £3.5 million 'High Street Fund', the resources for which had been scraped from the open palms of such incorrigible givers as Barclays, BP, Capita, Deloitte, RBS and Lloyds. For every smashed window on an outer city high street in Haringey or Charlton, a Mother Theresa from the auditing-accountancy oligopoly; for every small proprietor grieving over his emptied shelves, a battery of policy documents bearing solemn promises of *maximum regeneration impact*. Thus the wounds of our broken society are cauterised. But for those whose abundant faith in universal human benevolence and Deloitte is tempered by intelligent pragmatism, questions impudently arose. Who would administer these funds? Who would be granted the honour of bean counting his way back to social stability? No answers were immediately forthcoming. As the days ticked by, and as the reports on judicial repression poured in, and as retailers of plate glass watched as their sales figures boomed, still the questions went unanswered. Six weeks passed. Prisons bulged. At last, on 4 October, the Mayor's Office announced that the Greater London Council's two 'high-powered taskforces' would be led by Sir Stuart Lipton (target: Tottenham) and the merely plebeian Julian Metcalfe (target: Croydon). They are a symbolically significant pair, this corporate Tweedledum and Tweedledee, for Lipton is the real estate developer responsible for the City's grim financial services ziggurat, the Broadgate Centre, while Metcalfe founded Pret A Manger, under the proliferating striplighting of which migrants from the EU are today forced to beam inanely, for a subsistence wage, in satisfaction of the all-too-effective demand from high-wage FIRE sector workers for luxury afternoon comestibles. Here they are, Lipton and Metcalfe, one responsible for the palace of the financial services sector, the other for its 'ethical' servants' barracks, the torn halves of a whole which in fact *does* add up, like the pieces of a simpleton's jigsaw, to provide an image, in charming miniature, of our domestic model of economic 'growth'.

The first draft of the following text was written before this was yet known. It was, or was meant to be, an early and desperate attempt to leap off the carousel of cretinous recrimination on which fascistic journalists, 'left wing' politicos and automated politicians all rode, weeping and gurning and clinging to their war ponies. As such, its sections constitute a preliminary investigation not into deprivation or senselessness, nor yet into 'austerity', but into the new profitable industries in immiseration that have for the last few years *flourished* in what establishment discourse insists is a void; one which is impossible to remedy unless Sir Stuart Lipton and Julian Metcalfe expand into it. It doesn't take an unreconstructed clairvoyant to guess that Johnson's extra £24 million will be dynamically pissed away on the miserable step-changes of a regeneration policy from which no one save Pret A Manger's new owners – Goldman Sachs – are apt to benefit: and still there is work to be done.

ORGANISED CRIMINALITY I
(AKA UNEMPLOYMENT)

For a while now we've been inundated with phenomenologies of the new regimes of workfare, developed by new graduates whose abandonment to the desiccated labour market at least affords the opportunity to put their Husserl to good use. But, some questions: what about the macroeconomics of unemployment in the UK? How can we understand the misery of work in terms of the misery of worklessness, when these two conditions are becoming ever more intimately related in the lives of most proletarians forced to live in the dungeon of domestic capital accumulation? We ask not only about the degradation of conditions of employment down the supply chain, in the informal sectors of the international labour market (even fascistic fashion journalists allow their moral fibre to bristle at the thought of those kinds of globalised immiseration), but about the conditions of life for those who 'succeed' in climbing out of unemployment and into work.

The question first of all demands the iteration of some basic banalities. Like all good communists, we repudiate without hesitation the reactionary separation of *consumerism* and *production*, the privileging of the former over the latter, and the attendant pantomime dreamwork of consumer metaphysics, which tendentiously positions all humanity in an income continuum which denies all *qualitative* distinctions, including especially (and intentionally) the distinctions of class, whose truth outvies the corruptions of mere accountancy. Now more than ever the interface of 'work' and 'consumerism' in our society is rotten: it is the loop by which long term structural unemployment recreates the market for low end consumer commodities and by that

means also *recreates the jobs which the long-term unemployed are expected to aspire to.*

In Peckham, the windows of Burger King are smashed in. In the cavernous Lidl next door to it, the five members of staff who at any one time provide *the entire workforce* are pulling down the shutters. The manager is probably grateful for the break, because now he can get the cashier to do some sweeping. In Lidl labour is many-sided: exhaustion and humiliation have a great number of facets. The worthlessness of the majority of the jobs available in an economy whose service sector has expanded is the qualitative critique of unemployment statistics.[1] Companies like Lidl, Aldi, Primark and Fortune 21, celebrated in the Companies sections of the Quality Financial Press for their swift footwork in the face of market headwinds – i.e., for profiting from new poverty – have expanded rapidly over the last two years. Lidl and Aldi increased their market share by 10 percent in the three months running up to January 2011.[2]

All this would be old news if only it didn't dovetail so nicely with State employment policy. Capital management ideology states that because the administration of the unemployed requires the expenditure of taxation monies, long-term structural unemployment is a decelerator on private sector dynamism and nothing else. Naturally this is just a straightforward lie. The 'no frills' service industries are an important destination for the monies disbursed to benefit recipients. As the State continues to huff credit back into the financial sector, crossing its fingers and hoping that its preferred bubbles might reflate, it also endorses as never before long-term structural unemployment; and it therefore ensures continued and growing demand for the cut price commodities that the 'no frills' service sector provide and which the immiserated must consume. As anyone can see, this means a net transfer of jobs from the 'frills' service sector (i.e., old misery) into Lidl, Aldi, Fortune 21, Primark, and so on; and because the frills that these companies remove are, in the first instance, wages for their staff, the managed production of unemployment is equivalent to the managed degradation of the worst 'legitimate' forms of capitalist work. In this pirouette, bad work, which, as anyone who has ever replenished a shelf in a Tesco can attest, is not exactly pastoral (it is preternaturally shit), reaches a new nadir. Working class unemployment and working class work are remodelled together in a grisly bump'n'grind, and both are made ever more foreign to the forms of bourgeois work and bourgeois unemployment which the doctrinaire professors of bourgeoisdom – using their own class categories as a circular warrant – insist they must resemble.

While it is often pointed out that the sharp distinction between the public and private sector is specious – which is to say, an intellectual fabrication belonging to capitalists, whose next move is always to turn the distinction 90 degrees, so that it forms a hierarchy of value, i.e., private sector dynamism vs. public sector bureaucracy – the argument we make is slightly different. Because 'long-term structural unemployment' alters the structure of demand for basic commodities, which then need to be sold at a profit, it creates new impetus for the carnival of wage reduction – the race to the bottom as *spectacle* in the epoch of Silvio Berlusconi, eBooks and famine. In other words, 'long-term structural unemployment' is the etiolated hieroglyph of a new circuit of accumulation premised on the intensified exploitation of the employed. This is true even before we start to monitor State

efforts to push private-sector wages below the legal minimum which it imposes, as its civil service functionaries scramble to put to work their hypertrophied flexible skill sets in the innovation of new forms of 'workfare' torture.[3] As the civil servants might put it, the stagnation circuit is a 'formal' economy with all of the best features of the 'informal' economy (e.g., no guaranteed pay).

The State happily attaches to this profitable stagnation circuit a label marked 'aspiration', as if it were an expressionist ballet, which from the moderate heights of Downing Street it does perhaps resemble; and the press gesticulate at

---

judicial repression poured in, and retailers of plate glass watched as their sales figures boomed

---

their demonstration-model abyss and moan about the 'nihilism' of the riots which shattered the facades of so many betting shops; and the sky does not fall in. But fulmination against the destructive 'nihilism' of the riots merely underwrites what it claims to protest against, because it refuses to acknowledge just how much of our social 'fabric' would have to be destroyed if any social life were to be more than accidentally liveable. In the future, it must by now be obvious to anyone without a derivatives portfolio that effective social action will have to be *more and not less destructive*, because less and less of it is capable of being transformed merely by a modification in the terms of ownership. However important concerns about the 'proper' targeting of destruction may be, when their social function is to serve as a gag – and it usually is – better not to let them be put in your mouth.

HUMAN SUPERIORITY TO WONGA

In 2010, the online lender Wonga.com came eighth in the website Startups' list of the top 100 new companies. Providing reasons for Wonga's placement, the website states that

[CEO] Errol says [the company is] all about bringing a convenient and speedy solution to short-term cash needs, and lending as responsibly as possible. Wonga has already processed well over half a million loan applications and secured a $22m venture capital funding round in the summer of 2009.

This already says quite a lot. According to the Bank of England, in the last 25 years secured debt quadrupled as proportion of earnings; by contrast, unsecured debt rose only slightly.[4] This is all very useful for anyone who wants to reflect on property boosterism during the last round of British capitalist growth – even the economists who overpopulate the Bank of England's financial policy teams dimly intuit the truth of it – but the data also conceals some important areas for materialist inquiry.

Working class unemployment and working class work are remodelled together in a grisly bump 'n'grind

---

For example, it doesn't begin to tell us under what terms that unsecured debt was loaned out; nor does it tell us by whom – though the fact that economists at the Bank of England are uninterested in *this* is more understandable, since analysis of exploitation isn't one of the Bank's 'core purposes', any more than Ben Bernanke is the universal subject-object of

history, or any more than critical analysis of the organic composition of capital is one of the 'core purposes' of his maids. But, then, under what terms was the debt lent out; and by whom?

Between 2006-2011, as the retail arms of the banking sector put its orgies of ebriety behind it and got sensible, the 'payday loan' industry quadrupled.[5] What this means is that an increasing proportion of debt advanced to the working classes has been lent on increasingly severe terms. As debt support organisations wish so passionately to inform us, the growth in the payday loans industry means cycles of debt misery for the expanding industry's 'repeat customers'. Concerned parliamentarians have of course gladly seized the opportunity to emote some vote winning concern. In 2010, a noble band of 31 MPs clubbed together to sign a motion against payday loans lending. In the acrid mixture of intolerable circumlocutory jargon and decorous moralism so typical of parliamentary rhetoric, the motion stated that

the House [...] believes that lending of this kind can prove both socially and financially irresponsible and that the Government and appropriate regulatory authorities should insist on the application of the regulatory principle of fair treatment of consumers which currently applies to savings and investments in the UK to sub-prime lending products to protect vulnerable consumers.[6]

Shuffling aside the motion's gamut of illiteracies, we should note that it was essentially a call for *transparency*. Extortion is bad when it 'irresponsibly' uses 'television advertising' to delude oafish consumers – some of whom are vaguely 'vulnerable' – into acquiring debt which they won't be able to amortise. The focus

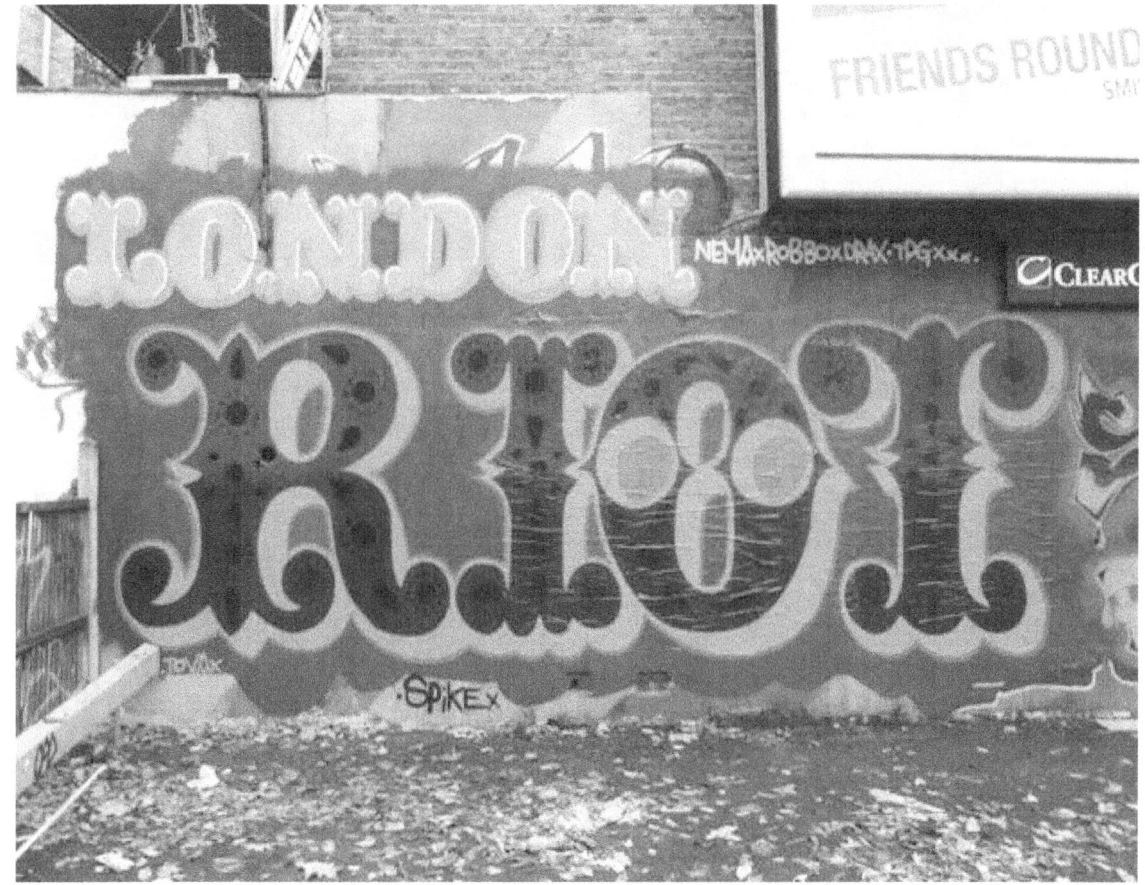

Eine, *London Riot*, graffiti on Hackney Road, London

on deception makes real social tendencies irrelevant since, according to the debauched enlightenment thinking of provincial MPs, social suffering is fine so long as it isn't advanced by mis-selling. 'Social causation' – that catch word of liberal determinism – is therefore transferred away from material reality and into the phantasmagoric graphic design hothouses of the maleficent advertising sector; and the nauseating vocabulary of consumer taxonomy (the 'vulnerable consumer', the 'satisfied consumer', the 'consumer in need', the 'thrifty consumer', the 'irresponsible consumer', etc.)

falsifies the two most significant consequences of the transfer of unsecured debt from the retail banking sector to the payday loans industry. Firstly it ignores the fact that more and more proletarian cash is canalised into debt servicing, the nearest equivalent to coal shovelling that the field of consumption has to offer. Secondly, it *occludes the involvement of conventional banking capital in the loanshark sector's growth*.

Who is protected in this conversion of 'irresponsible' lending into a *moral* problem? When politicians and liberal 'campaigners' pump their little fists and 'insist' on 'responsibility', what they do, whether they know it or not, is draw a *cordon sanitaire* around bad capital in the interests of good.

As Startups' lyric poem to Wonga suggests, in this sense squarely in the anti-physiocratic tradition of political economy, capital doesn't just grow out of the ground. Marxists should be more than interested in just taking the pulse of this growth industry: they should be climbing directly into its ventricles. The Chief Executive of Dollar Financial, Jeffrey Weiss, whose company (serving 'unbanked' and 'under-banked' consumers) is currently expanding into the UK, estimates that 'we're maybe 25 percent of the way towards a full country build-out... you extrapolate from our current 350 stores I think there is a potential universe for us of 1,200 locations.' And corporate tumescence on a scale this cosmic requires large sums of venture capital, at least some of which originates with *banks*, the newly discovered responsibility of whose retail divisions is what allowed the payday loans industry to jack up its growth forecast in the first place.[7] For example, Dollar Financial in 2009 emitted $600 million in unsecured notes (i.e., interest bearing debt) due for repayment by December 2016. Much of this debt is required to finance its aggressive expansion strategy (aka Jeffrey Weiss's macho 'build-out' on his 100 percent proletarian treadmill), but the extortion industries it is accumulating worldwide, including the Swedish exploiters Sefina and Folkia AS and the British pawnbroking business Sutten & Robertson, themselves rely on the official banking sector for their lines of capital credit. Dollar Financial's 'lines of capital credit' are the tentacles of 'legitimate' capital investment, seeking new returns.[8] Capital in the 'irresponsible' payday loans industry is *bank* capital or, indeed, is just *capital* as such, doing what it does best in a period when extortion

replaces asset inflation as the menacing guarantor of net valorisation.[9]

After smashing their way into the local money lending store, rioters on Peckham High Street had a shot at opening the safe. Since they weren't professional safebreakers, they finally contented themselves with smashing the counter and looting the chairs. When the police vans arrived, the chairs were hurled through their windscreens. If only Liberal Democrat politicians were less artificially horrified by the display of pure criminality – or whatever else is today's anti-analytic collocation of choice – they might commend its innovative lateral thinking and present the participants with Young Enterprise Awards, small cash prizes and an opportunity to visit the town hall to receive a framed A4 certificate from the mayor.

ORGANISED CRIMINALITY II (AKA CLASS FORMATION)

On Peckham High Street, on 8 August, as the riot moved up the road towards Camberwell, kids were pulling motorcyclists from their bikes. When one rider stood up and staggered in a daze back towards his vehicle, five or six men knocked him back over and kicked him. As his ribs were broken, the atmosphere in the street was festive. Anyone who didn't object was *in*: solidarity was extended to anyone who would by virtue of their presence underwrite the necessity of the action.

After the riots concluded with the obligatory surge of policing operations, and as a thousand bobbies flocked onto every street and into every KFC, these were not the experiences of community most discussed by the British bourgeoisie. When rasped by *its* lips, the term referred to any group of inscrutable 'others' who defended their shop windows or inventory, and who therefore – though they may not have meant to, or have said as much (but the British bourgeoisie can forgive their inarticulacy, because it takes such pleasure in speaking on their behalf) – *defended the sanctity of private property as such*. In the welter of blatant class strategising that ensued from 10 August, this use of 'community' was hatefully unavoidable. Across the land, talk show sofas buckled under the weight of garrulous 'community leaders', eager to retail the usage back to the bourgeois journalists who were principally responsible for its origination.

But because the definition of community as a property owning community is so truncated, banishing, as if by ASBO, a good number of forms of human sociality which might appear in fact to be highly communal, it requires some explanation; and class spokespeople wasted no time in stretching the meaning of another concept to fit its interests. Thus we were treated to a thousand slurred treatises on the concept of the 'gang'. The gang, we discovered, is distinguished from the 'community' on the grounds that a community defends private property whereas a 'gang' undermines it, or, if it doesn't undermine it, steals and eats it, or uses it to play Xbox.

This is of course, terrific bullshit. Upstanding 'communities' do not get to *have* the private property which they so valiantly defend except by benefiting from exploitation (whether directly or indirectly), or by dabbling in illegal forms of accumulation the profitability of which is to some degree tailored to level of risk. Bourgeois moralising of 'communities' vis-à-vis 'gangs' is therefore a means of deleting from its vocabulary of social explanation *the capital relation* or, in other words (and more simply), is a means of ignoring what people *have to do* in order to acquire the property which they later attempt to preserve,

But fulmination against the destructive 'nihilism' of the riots merely underwrites what it claims to protest against

and this is hardly surprising, since no one has got more from doing severely unpleasant things (to other people) than the bourgeoisie. And while there might certainly be 'exceptions' to the rule that property is usually acquired, in the first place, through exploitation or through crime (and who doesn't love an exception?), 'communities' which establish their small property holdings in the only other fashion possible, i.e., by submitting to the most hateful and never ending drudgery, are a source of moral inspiration only to those who enjoy the benefits of their servility.

Thus as the white movers and shakers of Dalston fall in love with the baseball bats which (they convince themselves) were wielded by Kurdish shop owners not, surely, on their own behalf but *disinterestedly*, or (the delusion is just as grandly disingenuous and really amounts to the same thing) on the behalf of the *community*, it isn't hard to point to contravening evidence. The website 'Gangs in London' claims that:

> [t]he families that make up the Turkish and Kurdish criminal organisations of London have imported 1,000's of kilo's of heroin into the United Kingdom during the last three decades. During which time they have made links with British criminals, including the so-called 'Liverpool Mafia', the 'Irish Godfather', Christy Kinahan, Albanian people trafficking networks, east London Asian crime families, Chinese Snakehead gangs and the predominantly black street gangs within the boroughs of Hackney and Haringey.[10]

To which list we could speculatively append Santander, Wachovia and a lustrous host of other cartels, owned predominantly by whites. There are no pure communities; there are only groups of people struggling to live under capital; and all the bourgeois weasel words of moral purity are at bottom a play for class hegemony,

dangerous both because they lend themselves to racism and because they can be put to work in stymieing any discussion which doesn't restrict itself to juggling a few moral platitudes siphoned from a discarded copy of *The Evening Standard*.

The quotation above therefore (and others like it) is useful as a preliminary inoculation against the worst stupidities of bourgeois response; but it evidently doesn't tell us much about the recycling of 'illegitimate' profits into 'legitimate' petit bourgeois enterprises – of the kind which splay along the high streets at the city's fringes, and which are now (or even now perhaps already *were*) the toast of those who would usually prefer to shop in outlets owned and managed by people of their own class and ethnicity. Nor does it tell us about the role of gang profits in accentuating ethnically profiled class differences within geographically defined 'communities'.

From our current outpost on the cloud of unknowing it would be irresponsible to speculate about different gang forms and, in consequence, it's difficult to speculate on the variant means by which gangs are plugged into different circuits of legitimate (low level) accumulation.[11] One valuable line of inquiry might investigate the historical relationship of gangs – and therefore of criminal profits – to the citizens' credit union, whose tightening of the sweet bonds of the *community* is extolled so lovingly by politicians eager to ensure that citizen-promoted debt servitude can replace traditional (redistributive) forms of state provision.[12] Red Tories, Blue Labour and the attendant gaggle of Christian citizen groups together, become the communitarian cheerleaders for forms of extortion and immiseration newly moralised because they are conducted on a 'horizontal' and not a 'vertical' axis.

CONCLUSION ON A HIGH

And when discussion comes to the issue of drugs – as discussion inevitably does, whenever it isn't on pause while the liberal discussants feign to weep about absentee fathers or rap music – the 'legitimacy' Hydra raises once again its terrible network of heads. Amid all the talk about the lamentable involvement of 'our youth' in the drug trade, with its descants on 'slow motion moral collapse' and its shrill undertone of bourgeois asepsis (i.e., just don't let them *infect* us), not much has been said about the involvement of the banking sector in 'handling' (i.e., laundering) the 'proceeds' of that trade.[13] It might just be conceivable that one or two of the people who last week lobbed bricks through the windows of their local Santander have, in the past, made some cash from street dealing, but it can be guaranteed that they didn't make as much as Santander itself, squarely implicated along with Bank of America, HSBC and others in large scale laundering.[14]

If laundered drugs money is a literally indispensable source of liquidity for a banking system gone costive on its own toxic debts and wallowing in 'public' opprobrium, it is obvious why these unacknowledged legislators of the world have *no interest* in even the managed decriminalisation of drugs. The task for us then is not to accuse the state of 'hypocrisy' – a term which belongs to art and bourgeois ideologues who, because they are removed from positions of political power, have no interest in disguising their prejudices – but to identify the contours of the complicity of state capital in nurturing organised crime; and to specify the dependency of legitimate (i.e., legally recognised) capitalist institutions on the profits generated in a purportedly 'extra-legal' economy.

In the last fortnight, no one could have failed to be subjected to riot analysis, unless they've had their head buried in the little pile of sand they keep specially prepared in their panic room. Such analysis is worked out usually by means of tidy acts of discrimination, accomplished at various degrees of sophistication or crudity, by salaried entities more or less human. By now we know the cast of the drama they construct: looters and till workers, citizens and politicians, organised gangs and anomic school kids, shopkeepers and crime lords, the deprived and the comfortable, those whose acts can be understood *if not condoned* and those whose privilege earns them only contempt. The distinctions have this character, that they eventually subordinate themselves to a more abstract distinction between victims and perpetrators. With the astigmatism of class hatred now so well implanted and the judicial

---

more and more proletarian cash is canalised into debt servicing

---

system running on autopilot, Justitia has no need to put her blindfold back on.

But it is necessary for capital and its representatives to avail themselves of all their resources of discrimination. The resources are a desperate bulwark against the apperception of what binds us together, which is the social necessity for capital to accumulate under conditions of straitened reproduction.

Time will continue to go by, and, most likely, it will continue to be said that gangs are the consequence of inner city deprivation. That statement has an implicit tendency. If only more capital would flow into the estates in which their members live, gangs would melt into

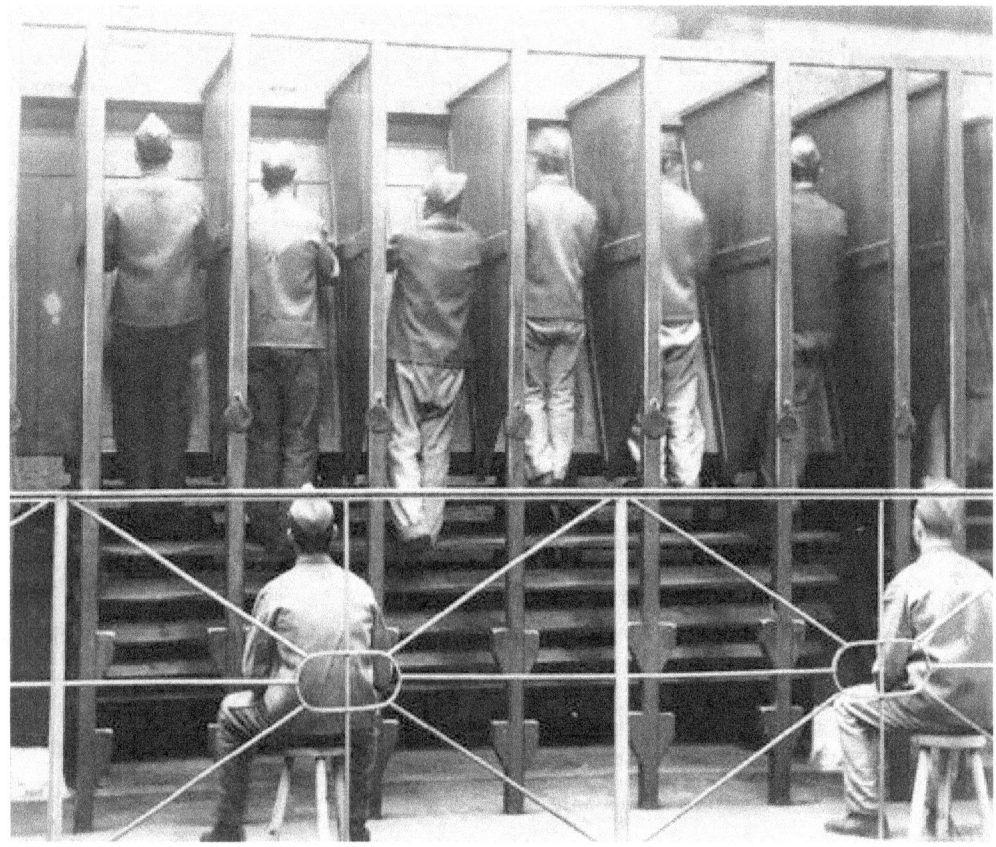

Prisoners on the treadwheel at Pentonville Prison, 1895

air. Just as in Tottenham the looters seemed, according to the standard colonialist clichés, to 'melt' into the side streets. Once again, we can feel safe in 'our' cities. This is the official, rubber-stamped mythos of capital accumulation. The gangs exist because their members are separate from capital. Where there is no capital, there is deprivation. This is deprivation's definition. The mythos serves two functions. First, by premising deprivation and its dark bedfellow, 'crime', on separation from capital, it apes the angel of history and announces capital's absolution, because capital cannot be responsible for that from which it is separated and which it no

doubt yearns in its heart to reach. Like Jesus, capital loves all human resources. At most (according to this argument), the fault lies with doltish town planners, callous politicians and other marionettes in the shop window of social obfuscation. Second, the mythos provides an abstract and tendentious 'vision' of what at any given point accumulation is, in the sense that it suggests that capital investment can be expanded at will to eradicate the stain that, e.g., the Pembury Estate has left on our community, but also, by parity of reasoning, in the sense that it implies the real circuits of accumulation do not currently depend for their completion on exactly the forms of human torture brought about by reified 'underinvestment', but instead just happen to operate elsewhere, in other places, far away from the sad and frightening little people who spoliate their local Betfred. This report has attempted urgently to specify the contours of a few of the new and profitable immiseration industries. It has done so not just because they exist to be pointed out, but because we hate the fatuous moral philosophies which take a professional interest in ignoring them. As the commentators continue to waddle

it must be obvious to anyone without a derivatives portfolio that effective social action will have to be more and not less destructive

onto the tattered red carpet under the 'public eye' to play with their Fisher Price scale of good and evil, performing narcissistically their games in moral philosophy, and refining their breviary of 'sociological' distinctions, who can say how many extra burnt out cars and shops and flats it will take, and how much more concentrated misery and desperation and venom, to demonstrate in spite of all this that in at least one respect we truly and undeniably are all in this together: under capital, right up to our necks.

FOOTNOTES

1   For a very useful worker's inquiry conducted by an employee of Lidl, see, Frank L. Ludwig's 'The Lidl Shop of Horrors: Working Conditions in Lidl', http://franklludwig.com/lidl.html. See

also, 'Arbeitsbedingungen bei Lidl', *Der Spiegel*, 10 December 2004, http://www.spiegel.de/wirtschaft/0,1518,332171,00.html

2   Both of these companies have flourished by the enthusiasm with which they strangle their workers. For example, Lidl's wage policy has always been based on the imposition of impossible workloads which must be completed by workers during (unpaid) overtime, which is to say, on the unpoliced and unremunerated extortion of extra-contractual labour.

3   Corporatewatch recently posted an interview with a 24-year-old Bangladeshi woman made to work three days a week in Primark as a condition for the receipt of her Jobseeker's Allowance. Assuming a work week of 21 hours and a subsistence 'benefit' of £50.95, this woman was paid an hourly wage of £2.43, (http://www.corporatewatch.org/?lid=4030). This is obviously a rather venerable means of manipulating the market rate for 'unskilled' labour, but as state politicians natter on about 'skill creation' through workfare, which is to say, as they continue to promote virtualised accumulation for the exploited, the real exigency for wage regulation reveals itself: competition is hotting up in the no frills sector. Last month US Primark clone Forever 21 opened its first retail branch on Oxford Street, http://www.ft.com/cms/s/0/a72a0814-b863-11e0-b62b-00144feabdc0.html#axzz1VChqrtoy This might, as the *Financial Times* boorishly reports, be good news for 'teen consumers', but price competition in low margin retail industries requires the licensing of new peaks in worker exploitation. Otherwise the margins go sour.

4   Matthew Hancock and Rob Wood, 'Household Secured Debt', *Bank of England Quarterly Bulletin*, Autumn 2004, http://www.bankofengland.co.uk/publications/quarterlybulletin/qb040302.pdf

5   For anyone who wants 'proof' that bank lending to individuals fell off from late 2008, the Bank of England provides a graph, http://www.bankofengland.co.uk/statistics/li/2011/Aug/chart6.gif

6   Early day motion 1194, http://www.parliament.uk/edm/2009-10/1194

7   See Rupert Neate, 'Loans start-up Wonga gets $22 boost from Facebook backer', *The Telegraph*, 8 June 2009, http://www.telegraph.co.uk/finance/newsbysector/banksandfinance/5476027/Loans-start-up-Wonga-gets-22m-boost-from-Facebook-backer.html Balderton Capital, the main supporter of Wonga, claims neatly enough that most of

its US$1.9bn capital comes from University endowments. Academics at Harvard would do well to reflect on this as they release themselves upon the tiny cucumber sandwiches served at the latest social justice colloquium.

8   http://www.reuters.com/finance/stocks/DLLR.O/ key-developments?pn=2 and http://www.reuters. com/finance/stocks/DLLR.O/key-developments/ article/2035785

9   For some (now somewhat outdated) information on the debt *collection* growth sector, see: http:// www.washingtonpost.com/wp-dyn/content/ article/2005/07/27/AR2005072702473.html The best survey on recent private sector dynamism in the UK incarceration industries is Clinical Wasteman, 'Unlimited Liability or Nothing to Lose', reprinted in this issue of *Mute*, pp.XX

10 See London Street Gang media resource's 'London's Mafia? History of Turkish & Kurdish Organised Crime in London', http://gangsinlondon.blogspot. com/2011/03/londons-mafia-history-of-turkish.html

11 Even the bourgeois social-justice-police- surveillance enterprise, The Centre for Social Justice, acknowledges in its 2011 'Dying to Belong' report that little has been done in the way of usable (i.e., repression facilitating) history from above. Thus the report hilariously frets that '[t]he MPS found 171 gangs operating in London and the Home Office estimate that there are 356 gang members in the Capital. This would mean around two people per gang, which would not, by the Home Office's own definition, constitute a gang.' (p.20) And since there are presumably at least *some* gangs with three members or upwards, it would seem to follow that according to the Home Office's own estimates there exist many gangs with *one member or fewer*, http://www.centreforsocialjustice.org.uk/ client/downloads/DyingtoBelongFullReport.pdf As the notorious gang Walt Whitman once put it: 'I contain multitudes'.

12 An *extreme* instance of the kind of informal credit network operated via gang formations is reported by David Cohen in 'Heroin wars, loan sharks and executions: the Turkish gangs terrorising north London', *The Evening Standard*, 17 November 2009, http://www.thisislondon.co.uk/standard/ article-23770522-heroin-wars-loan-sharks-and- executions-the-turkish-gangs-terrorising-north- london.do. Naturally any instance *less extreme* would bear less interest for journalists writing for this paper.

13 See Michael Smith, 'Banks financing Mexico gangs admitted in Wells Fargo deal', Bloomberg.com, 29 June 2010; http://www.bloomberg.com/news/2010- 06-29/banks-financing-mexico-s-drug-cartels- admitted-in-wells-fargo-s-u-s-deal.html

14 In an excellent recent essay in black-market/ legit-market partnership, John Barker notes that '[f]igures extrapolated by [Tom] Feiling suggest that just one percent of the retail price of cocaine in the USA goes to the Colombian coca farmer; four percent to its processors and 20 percent to its smugglers. Seventy-five percent therefore is realised in the rich world.' How much of that is realised *as profits* by the rich world's banks remains to be assessed. See 'From Coca to Capital: Free Trade Cocaine' in *Mute* Vol3 #2 Spring/Summer 2011, http://www.metamute.org/ en/articles/from_coca_to_capital_free_trade_ cocaine_0

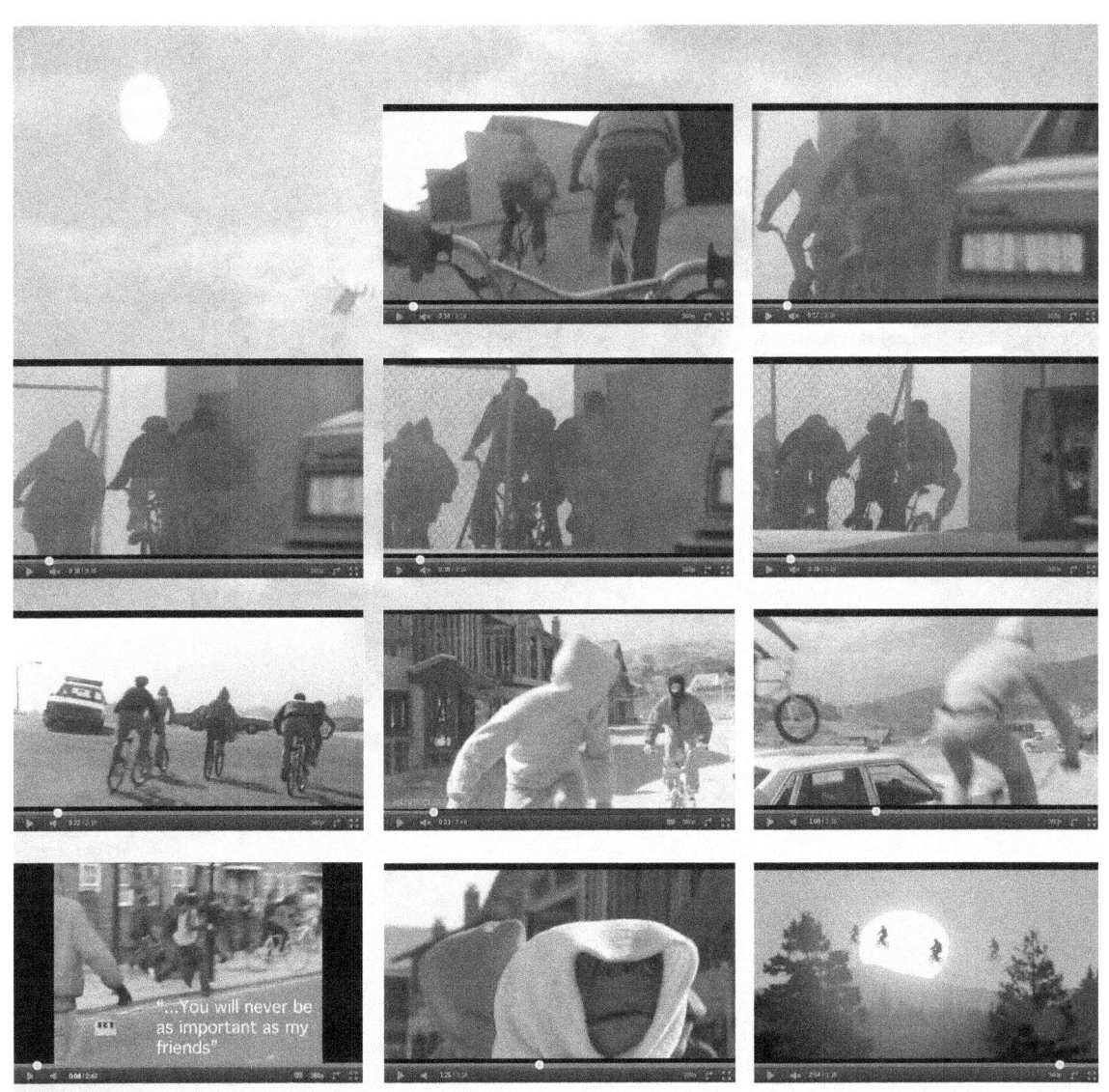

Rachel Baker, *Alienation Affect*, 2011

THE DARK
ARTS

Gregory Sholette's book, **Dark Matter***, provides a useful collectivising term for those artists who produce the art world from below. But, wonders* STEFAN SZCZELKUN*, how can we talk about cultural exclusion without thinking seriously about class?*

When the excluded are made visible, when they demand visibility, it is always ultimately a matter of politics and a rethinking of history. This is often the case with artists collectives.

– Gregory Sholette

The inventiveness of the everyday, the commonplace, and the nondescript multitude. In the age of deregulated aesthetic practice such dark matter inevitably intervenes within the valorisation process of official artistic production.[1]

The 'dark matter' in the title of Gregory Sholette's book refers to all the human creativity that is excluded from the mainstream art world. The book is a dense interweaving of political contexts, theory, accounts of radical art activity and considerations of the archive, mainly in a US context. The structure of the book is loosely related to Sholette's involvement in a series of political manoeuvres – the first of which is the Political Art Documentation/Distribution group that was set up after a call from Lucy R. Lippard in 1980. The PAD/D collection was donated to New York's Museum of Modern Art (MoMA) archive in 1989. For its author, the subversive implications of this donation provides a recurrent fascination throughout the book.

Following his account of this earlier history is a chapter based on Sholette's involvement in a group called REPOhistory in New York. Initiated with a series of triangles put up in public in 1994 to mark the death of Marsha P. Johnson, a leading LGBT activist, it subsequently led to the erection of a series of nine similar triangles under the project Queer Spaces.

Sholette then discusses the over-abundance of signs, creativity and collections heightened by our present level of commodity accumulation. Two artists' groups that worked critically in this area were Public Collectors and Temporary Services. Public Collectors recognise and bring attention to:

Temporary Services 'seeks to generate a non-market, non-accumulative economy of generosity.'[2] The account of their Chicago based project of 2000, *Free for All*, reminds me of the Free Stuff Parties organised by UK artist Mark Pawson at around the same time. This section ends with a discussion of the more confrontational actions of Etcetera in Argentina in 2002: the anti-BP actions of Liberate Tate in London in 2010 and Yomango's surrealist theft actions from 2003 in Barcelona, Spain, which seem especially relevant now in the light of the recent rioting in Britain. The collective gesture of Yomango, which means simply 'I steal' in Spanish, clarifies the symbolic dimension of looting.

Sholette goes on to examine Critical Art Ensemble as one of the leading 'tactical media' groups of the last decade. Sholette himself was involved in supporting the defence of CAE member, Steve Kurtz, who was charged with biochemical terrorism following 9/11. Kurtz was finally acquitted in 2008 after a long, hard court battle, but Sholette does not discuss his own support role here. He claims that CAE's artwork, *Molecular Food Invasion*, used the art gallery as a platform for public discussion.

Finally a chapter entitled 'Mockstitutions' on organisational forms and mock institutions focuses on The Yes Men and other artists' groups. This includes some original survey research into 67 of these groups, which I'll come back to.

A recurrent theme in *Dark Matter* is that of subverting dominant cultural values from within the archives. What do the mainstream museum archives like MoMA's gain from taking records of artists' dissent into their care? Sholette's idea is that these stored materials are 'a mark or bruise within the body of high art. The system must wear this mark of difference in order to legitimate its very dominance. Absolute exclusion is out of the question.' Sholette suggests that these ingested materials will undermine the system and what starts with a 'bruise' can lead to an 'infection' of the whole body:

> The perforation of a once suppressed archive exposes the wounds of political exclusions, redundancies, and other repeated acts of blockage that wholly or partially shape this emerging sphere of dark-matter social production.[3]

London's Tate Gallery archive has a mission that is informed by the scale of its immense funding base and position as the national flagship museum of contemporary

Dark matter is becoming lighter

art. It assumes that it is best placed to interpret radical art history. Who else could take proper care of the archived originals? Who else could attract such esteemed and learned writers to interpret this material and produce a reliably normative view of the past? What they 'forget' is that, as an organ of the State, Tate is of necessity embedded in its ideological apparatuses, self-serving rituals and practices. At the end of the day, it is fundamentally unable to critique its paymaster – the system – except by way of mild chastisement or warning, so that the Leviathan can adapt, modernise and survive. Not only that but the difficulties of access that result from professional archive practices mean that the radical cognoscenti themselves find it hard to get their filthy hands on the material. Once ensconced in such a vault, radical material is unlikely to be activated in the revolutionary, or at least, playful and irreverent way that was often intended by the original sources. Sholette, however, takes a more optimistic viewpoint and states that:

> These bits and pieces of generalised dissent resist easy visualisation, forming instead a murky submarine world of affects, ideas, histories, and technologies that shift in and out of visibility like a half submerged reef.[4]

In my experience, art world institutions will tend to intuitively reject anything which could undermine their wellbeing, or only accept an inoculating dose of such material.[5] But Sholette allows the 'dark matter' that has sneaked through these portals much more power:

> The archive is consuming its host, brandishing all the malicious resentment of the profaned, the philistine, the exile.

> A materialising dark matter now confronts this so-called future as a grinning archive and antagonistic corpse.

> [dark matter] directs our attention towards an ellipsis within the historical record where none is supposed to exist.

> The archive has split open.[6]

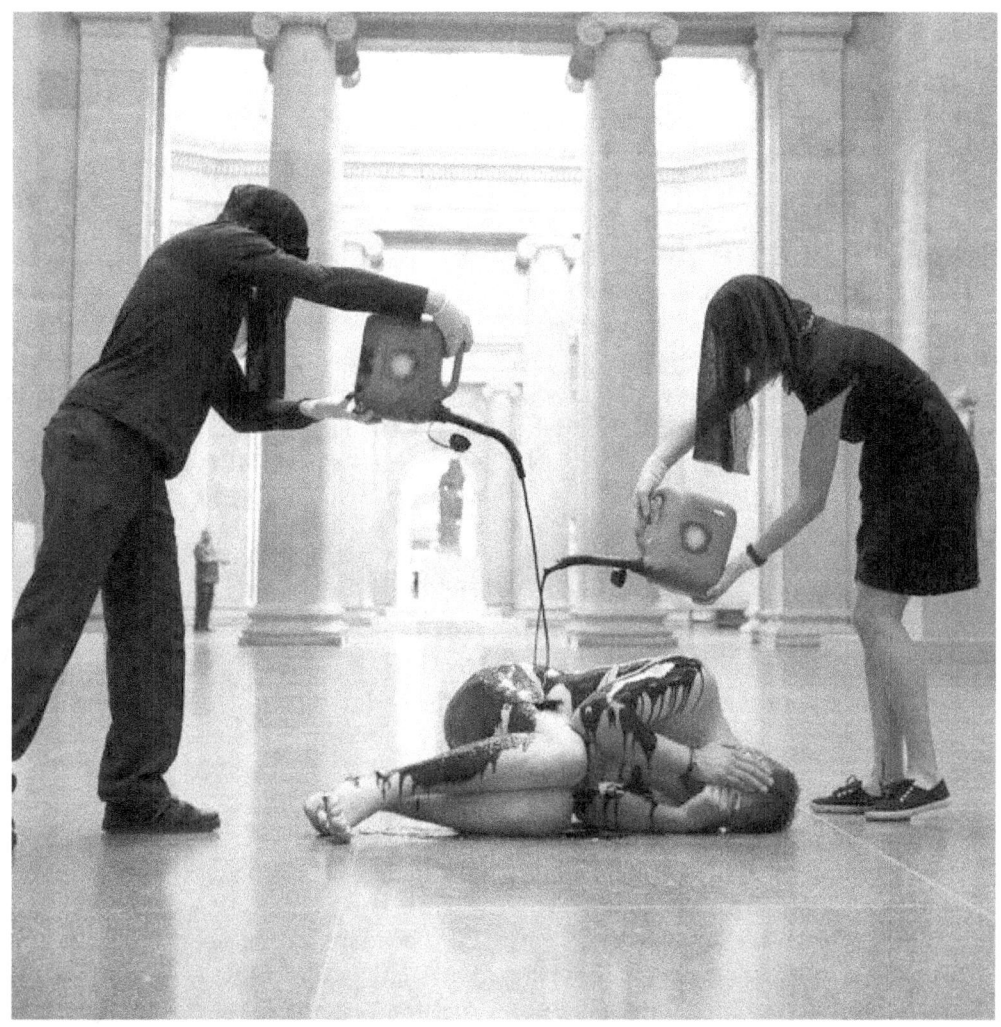

Liberate Tate's action marking the first anniversary of the oil spill in the
Gulf of Mexico at Tate Britain, 21 April 2011

DANGEROUS MATERIAL

Is PAD/D dangerous material? Or is it enclosed, de-activated, isolated and neutralised material? Is not the stuff donated to these vaults only given for lack of any other more open activist-run alternative? What we need is more in-depth and independent studies of initiatives that put their material firmly and independently into the public domain. At the same time we need to demand better access to, and digital publication of, collections that the State would probably rather keep buried. This means a lot more clamour needs to be made about the importance of initiatives such as artists' collectives.

I should now admit that I have long pursued an engagement with dark matter, mostly of different genres to those considered in this book – mostly relating to larger collectives. In the last 15 years I have even tried to drag these things into the light of formal knowledges: from my 2002 doctoral study of Exploding Cinema at the Royal College of Art to my current active archiving, with many others, of the Brixton Artists Collective, 1983-86, as *Brixton Calling!*[7] The material traces of which are, in spite of my reservations, heading to the Tate archive.[8]

Whilst doing the Exploding Cinema research, I became aware that the British Film Institute's (BFI) London collection was thin with regard to amateur home movies on standard and Super 8, the technology that took filming to the masses from the '60s onwards. When I approached them about housing archive material from Exploding Cinema, I was told that they did not take VHS tapes as they weren't of professional quality. Of the 1,000 or more film-makers that had shown films in Exploding Cinema in the pre-digital period of my study, many left copies of their works with collective members on VHS cassettes. So, inflexible

adherence to that rule meant that the films shown at Exploding Cinema are by and large not archived and likely to be lost to posterity. Whilst such rules are ostensibly about objective professional standards and so on, in fact they act as a filter on dark matter – in this case the zero budget self-made films of the 1990s. My study of the collective and its material traces and events are now archived but the actual works shown, the heart of what Exploding Cinema was about, are not included in the national archive of the moving image. My overall experience is that very little material from radical artists' collectives is in public archives and what *is* there is hidden by unfriendly index terminology or other means.

CLASS AND EXCLUSION

Sholette sees the new explosion of dark matter's visibility via the internet as characterised by:

a lack of interest in abstraction coupled with a fondness for everything that was once considered inferior, low and discardable. Qualities that were anathema to modernist notions of serious art.[9]

He thinks that digital technologies have put much informal social production into the public sphere. Frustratingly, he does not theorise the relation of 'low' material and 'informal' production to class and in fact, at one point seems explicitly to reject a class analysis.

Tactical Media is [...] a rejection of most nineteenth and early twentieth-century leftist movements and of the idea that the working class is a unique and ontologically privileged force of social and historical transformation.[10]

On the other hand, he thinks that tactical

media groups have emerged from a situation of 'surplus talent' and a what he calls, 'prickly, working-class imagination'. He doesn't go on to detail what he means by this prickly quality of imagination; instead, he goes back to a roll call of '60s counter-cultural influences and somewhat loses his way in the process of considering 'right wing' dark matter.

PRICKLY FILTERS

In another prematurely aborted theoretical analysis, Sholette discusses the concept of the public sphere through Grant Kester's dialogical aesthetics.[11] He refers to Oskar Negt and Alexander Kluge's concept of a 'counter-public sphere' suggesting that a discourse on dark matter would mean, 'constructing filters contrary to those of the market.'[12] A good point, and it's a pity there is no further discussion of what such 'prickly' friendly filters might look like.

One of the most interesting ideas the book puts forward is that dark matter, although deprecated, is essential to high culture. He claims that it plays 'an essential role in the symbolic economy of art.'[13] This is a claim that reappears through the book:

> What is not recognised, what cannot be admitted by the maintenance crews of the high culture industry, is the degree to which not only the art world's imaginary but also its economy is stabilised by the invisible labour of this far larger shadow economy. The material and symbolic sides of these economies endlessly amplify each other.[14]

He makes the point that all our informal conversations about figurehead art institutions help to maintain and reinforce their credibility and power. I feel this dependence of the art world

What do mainstream museum archives gain from taking records of artists' dissent into their care?

on dark matter is true, in the same way that the owning class has a dependence on labour (which is magically reversed in their ideology). However, I don't find Sholette provides the evidence to take this argument forward to the point that it could be used convincingly to argue for a radical redistribution of resources within culture.[15]

ART GLUT

A set of unobtrusive art world institutions and practices by art critics, historians, collectors, dealers and administrators have 'inscribed' the antagonism shown by early modern artists and are themselves increasingly programmed, as if on rails, with little critical self-awareness. This results in an art world that is 'afraid to admit that it is comatose'.

The book considers the mystery of the 'glut' of artists; 'the normal condition of the art market', as Carol Duncan pointed out in 1983.[16] Sholette provides some useful information on this overproduction. For example, in the USA between 1970 and 1990 the number of artists doubled. In the EU, the numbers of artists is the same as the working populations of Eire and Greece. In London, the number of self-described artists comes second only to the number of people employed in the city's business sector.[17]

Drawing on these insights, Sholette poses the following evergreen questions:

1. If this glut is commonplace and enduring, then what 'material benefit' does the art world get from this 'redundant workforce'?
2. Inequality between artists' incomes has increased in recent years; what would it take to politicise these excluded artists? And what action might it lead to?

3. Is self-organisation an effective counter to the 'exclusionary mechanisms of the art market'?[18]

Sholette has no answers to these questions and only comes to the conclusion that 'artists gain improved social legitimacy within the neoliberal economy while capital gains a profitable cultural paradigm in which to promote a new work ethic of creativity and personal risk taking'.[19] Artists currently voluntarily contribute to the symbolic system, even though they must be aware that, in most cases, it guarantees their own 'failure'.[20] Later, he calls this a 'managed system of political underdevelopment'.[21]

GROUP STIGMA

His survey of artists' groups leads him to draw two conclusions. Firstly, in spite of the efforts of the management theorists, there is still an art world stigma against 'multiple authorship', or presenting as a collective or group, although this is decreasing. This stigma may be mainly due to group membership usually being in flux; regular self-terminations happen for all kinds of reasons including internal disputes, resignations of key members, or even commercial success. Sholette's survey shows that groups are (or appear to be) marginal players in the art world.[22] Secondly, in the 1960s and 1970s groups were taking on organisational forms that mirrored the structure of the organisations that provided them with financial support. This has shifted since the 1980s and there is now less interest in organisational conformity:

Perfunctory compliance with official cultural regulators may have been a sporadic though unspoken practice by artist groups in the past,

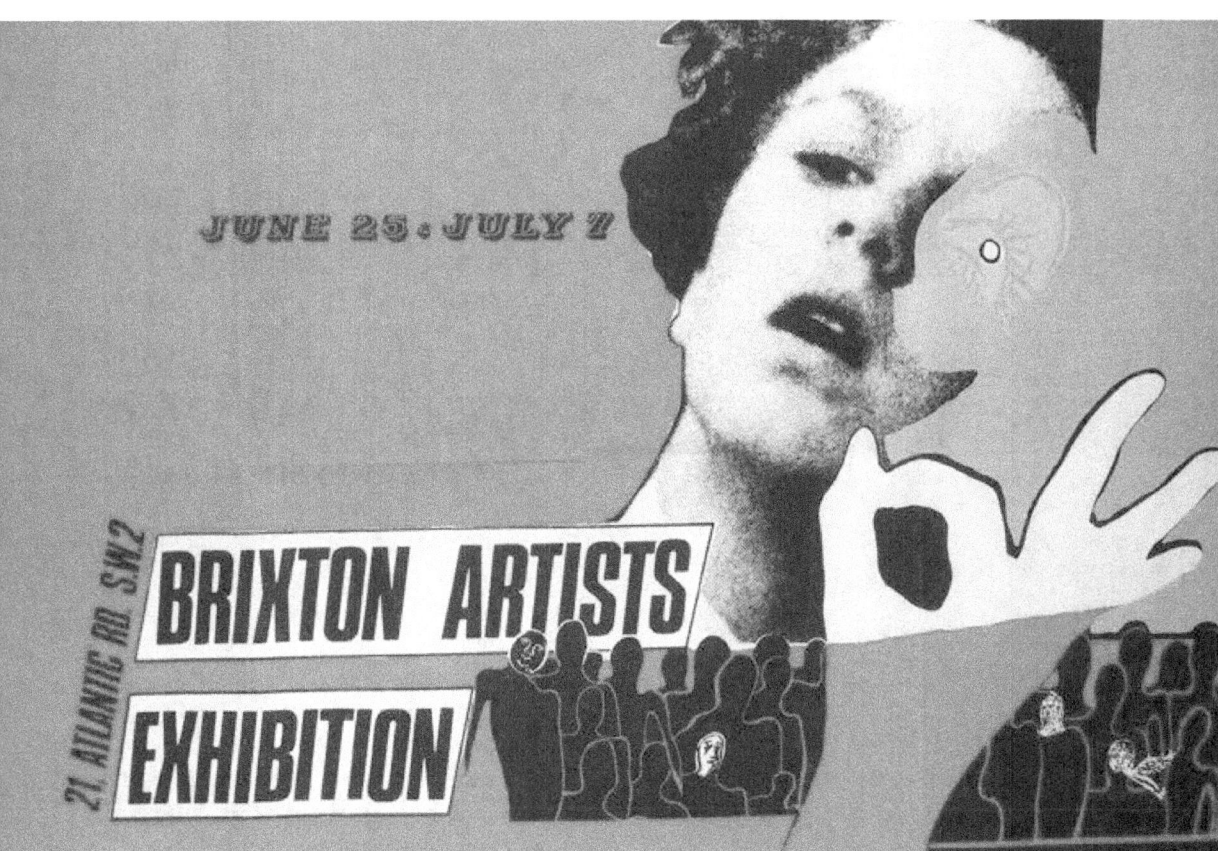

Publicity poster for Brixton Artists Exhibition, Atlantic Road, 25 June – 7 July, 1983

but today in an age of deregulation and semiotic warfare, such tactics are becoming pervasive, even amongst groups who, at least on paper, appear to be commercial enterprises.[23]

This is a mildly encouraging sign, but, due to the lack of ontological refinement of the differences between types of artists' groups, Sholette's conclusions about the nature of collectives as typical dark matter seem weak. In particular, there is no distinction drawn between a group of two or three people working together and larger, open collectives with internally democratic structures.

In terms of defining collectives as more or less autonomous, more or less radical, and more or less dark matter, the two most important things, as far as I am concerned, are these: the extent to which the collective is permeable to oral culture and working class people and the extent of independence it can maintain from commercial and state run institutions. Several of the larger collectives in my experience had an open door to new members – the currency of membership was the labour put into the project. They also had a clear no selection policy when it came to what work was shown. 'NO STARS, NO SELECTION, NO TASTE' was one Exploding

ingested materials will undermine the system, what starts with a 'bruise' can lead to an 'infection' of the whole body

Cinema slogan. Brixton Artists Collective 1983-86 decided upon shows at open public meetings. Membership was open to all for a few pounds. Both these collectives had a certain style in their spaces that was orientated towards the general public rather than an elite art audience. This kind of thing is anathema to the art world, where selection and the creation of exclusivity and the celebration of the possession of elite skills and knowledge is their raison d'être.[24]

CONCLUSIONS

What is really needed in this book is a class conscious theory of how culture works. The use of the categories 'formal' and 'informal' is unsatisfyingly vague. I'd prefer to use the historical separation of literary culture from oral culture (in Europe) as the dynamic that drives dark matter away from the light of publicity. This could help bring some clarity to the question of class.

I'm using 'oral culture' here in a quite specific way, one which I've been reminded by *Mute* editors is not yet in common parlance. The meaning I like to inflect oral culture with is a global one that encompasses all the processes by which we come to agreements on meanings, or by which we beg to differ, from time immemorial. As the zone of direct and relatively unmediated communications, oral culture is the main forge of human language and I don't believe, for instance, that Shakespeare invented the word 'bubbles'. Oral culture is the seat of judgement for the plethora of meanings that soak our environment. Oral culture is the pool in which all the sense media of our communication thrive.[25] Any new meanings have to be taken on within these fluid and common spaces of communication. Just like the production of goods, the production of cultural meanings depends on the multitude.

Literary culture is of relatively recent origin, and, from where I sit, expresses a localised European geography that can be traced back a thousand years to the humanist break from the writings of the Christian monks, using their discovery of Arabic translations of the ancient classics (and particularly Aristotle). This formation, embedded in the early city states, received a tremendous boost with the invention of printing with moveable type in 1450. The whole publishing apparatus became a means to power for the new bourgeois class who usurped the publicity of the aristocratic spectacle with their own literary discourses and knowledge.[26] Literary culture became closely identified with bourgeois being. They still hold on to it as their birthright. The bourgeoisie tries desperately to control meanings and communication, to police oral culture, to put us each in a mind cage. But this domination is ultimately an illusion. The meanings we hold in common are, at the end of the day, decided and reaffirmed in the oral realm. This means there is tremendous power in attending to it, and how to direct it.[27]

Sholette seems to veer from a passionate belief in the disturbing, if not revolutionary, power of dark matter: 'self-organised dark matter inserting itself into the ripped fabric of neoliberal cities, from below', to ambivalent feelings that perhaps these practices 'subvert, and yet reinstate'. He seems unsure if these 'emerging aesthetics of resistance' are any more than 'tepid acts of delinquency or even bitter gestures of discontent'. He hopes that they at least provide 'an expectation' – but of what?

This is really a diary or compilation of his efforts, thoughts and various involvements, and I think I had hoped for his own subjective engagements to be more explicit and less academicised. The reason for this focus on a very

particular stratum of dark matter would then be more organic and less arbitrary. A global study of dark matter would take the kind of team effort and resources that go into compiling a major dictionary or encyclopaedia. The key question may be how any such institution could maintain its revolutionary integrity and class consciousness whilst carrying out such a task.

Sholette has invented a useful term that might well be taken up, and gives us a sporadic view of resistance through political art, but his style of authorship in *Dark Matter* is too conventional. I feel the style on the whole mutes his own analysis to occasional whispers, rather than making oppression and its exclusions the key definer of the 'from below' of cultural production. Cultural resistance is no new thing.[28]

Sholette thinks that 'dark matter is getting brighter.'[29] This may simply be a function of media technologies making all kinds of knowledge more visible, or it may signify a huge groundswell of demand for more democratic societies. Or, perhaps, these are two sides of the will to power in the oral realm, the struggle from below. This book certainly allows us to give a name to, and begin to focus on, the creativity and cultural resistance that exists outside the art world proper. It may be flawed and partial, but it represents a good start towards developing a discourse that I think needs to embed itself outside of the academy – within the fields of dark matter itself.

Stefan Szczelkun <szczels@ukonline.co.uk> is an artist, living in South London, with an interest in open artists' collectives and networks

INFO

Gregory Sholette, *Dark Matter: Art and Politics in the Age of Enterprise Culture*, London: Pluto Press, 2011.

Links to Greg Sholette's strata of dark matter:

Candida Television: http://candida.omweb.org/
Journal of Aesthetics and Protest: http://www.joaap.org/
6+: http://www.6plus.org/borcila.html
Howling Mob Society http://www.howlingmobsociety.org/
Critical Spatial Practice: http://criticalspatialpractice.blogspot.com/
MicroRevolt: http://orangeworks.blogspot.com/
Center for Tactical Magic: http://www.tacticalmagic.org/
Yomango!: http://yomango.net/ and http://www.yomangoteam.com/
The Yes Men: http://theyesmen.org/
Critical Art Ensemble http://www.critical-art.net/
Target Autonopop: http://www.targetautonopop.org/
Temporary Services: http://temporaryservices.org/
including their wonderful Public Phenomenon Archive
see also http://www.darkmatterarchives.net/?page_id=21

FOOTNOTES

1 Gregory Sholette, *Dark Matter: Art and Politics in the Age of Enterprise Culture*, London: Pluto Press, 2011, p.99.
2 Ibid, p.100.
3 Ibid, p.11.
4 Ibid, p.34.
5 There are of course always exceptions to what is accepted. It can come down to personalities. When Simon Ford worked at the National Art Library he bought in a load of scurrilous zines and mail art. He never seemed intimidated by the institution whatsoever. Further back, Meg Duff at the Tate Library was very approachable and openminded, which might have been due largely to the liberal left support of ARLIS, a very well organised art librarians' union, which both Simon and Meg were leading members of.
6 Ibid, p.40, p.185, p.186 and p.188.
7 The original materials I collected are held in the BFI Special Collections and at http://www.stefan-szczelkun.org.uk/index2.htm
8 The alternative to the Tate's Archive is Andrew Hurman's self-funded and panoramic web archive of Brixton Gallery 1983–86. http://www.brixton50.org

9 Sholette, op. cit., p.30.
10 Ibid, p.145.
11 Sholette, op. cit., p.168. Grant Kester, *Conversation Pieces: Community and Conversation in Modern Art*, Berkeley: University of California Press, 2004. See also his *The One and the Many: Agency and Identity in Contemporary Art*, Durham, NC: Duke University Press, 2010.
12 Oskar Negt and Alexander Kluge, *Public Sphere and Experience: Toward an Analysis of the Bourgeois and Proletarian Public Sphere*, 1993, Minneapolis: University of Minnesota Press. This was a response to the perceived elitism or at least limitations of Jürgen Habermas' notion of the public sphere. Habermas does indeed privilege the literary over sense media in what I have read of his work. See footnote 20 below and Jürgen Harbemas, *Theory of Communicative Action*, 1981. The quote is from Negt and Kluge, ibid, p.188.
13 Ibid, p.3.
14 Ibid, p.44 and p.122.
15 For a classic exploration of art as collective action see, Howard S. Becker, *Art Worlds*, Berkeley: University of California Press, 1982.
16 Carol Duncan, 'Who Rules the Art World?' in *Aesthetics of Power: Essays in Critical Art History*, Cambridge: Cambridge University Press, 1983, p.172.
17 Sholette, op. cit., p.126.
18 Ibid, p.116.
19 Ibid, p.117. *Dark Matter* is good on how creativity and collaboration have become part of the new liberal management speak from Tom Peters to John Howkins. Artists who work collectively are seen by these people to have powerful problem solving skills, in contrast to the individualistic way in which art discourses have previously upheld genius and feared collective authorship. Perversely, it is by these business theorists that the idea that culture is inherently collective is being most successfully promoted.
20 I'd prefer a blunt talk about artist's oppression. See my self-published pamphlet 'Artists Liberation', 1986.
21 Sholette, op. cit., p.120.
22 Something that is very different in other media such as music and theatre.
23 Sholette, op. cit., p.163.
24 An 18-minute long oral history documentary on Brixton Artists Collective is available on DVD from 198 Contemporary Arts and Learning as part of it Brixton Calling! Archive show 29 October to 17 December 2011.

25 Let's not get distracted by things like all the
    working class people in literary studies, or, more
    widely, all those commoners who have gone
    through university in the last 40 years, or even
    by the achievement of mass literacy in the last
    century. Let's not be confused by the oral aspects
    of bourgeois culture. This is a theoretical idea
    somewhat like the distinction between lifeworld
    and system.

26 This history can be followed in Jürgen Habermas,
    *The Structural Transformation of the Public Sphere:
    An Inquiry into a Category of Bourgeois Society*,
    Cambridge: Polity (1962 trans. 1989) and can
    be read alongside Lucien Febvre and Henri-Jean
    Martin, *The Coming of the Book: the impact of
    Printing 1450 - 1800*, translated by David Gerard,
    London: New Left Books, 1976.

27 I might owe a debt here to Hannah Arendt's idea of
    power. I have to admit to reading Arendt indirectly,
    through e.g. Jonathan Schell, *The Unconquerable
    World: Power, Nonviolence, and the Will of the
    People*, New York: Metropolitan Books, 2003. http://
    stefan-szczelkun.blogspot.com/2010_02

28 E.P. Thompson's *Customs in Common* and work on
    the construction of 'folk' music by people like Dave
    Harker and Bob Pegg in the 1980s, showed that
    cultural resistance was always a response to the
    imposition of power. Another heroic effort, within
    a very different stratum of dark matter in view,
    was made by Emmanuel Cooper in *People's Art:
    Working Class Art from 1750 to the Present Day*,
    Edinburgh: Mainstream Publishing, 1994. We could
    even trace the idea back to Jean Dubuffet's use of
    the term Art Brut, or 'Outsider Art', as it came to
    be known in the English speaking world thanks to
    Roger Cardinal's work of 1972. See, Roger Cardinal,
    Outsider Art, London: Studio Vista and New York:
    Praeger Publishers, 1972.

29 Sholette, op. cit., p.3.

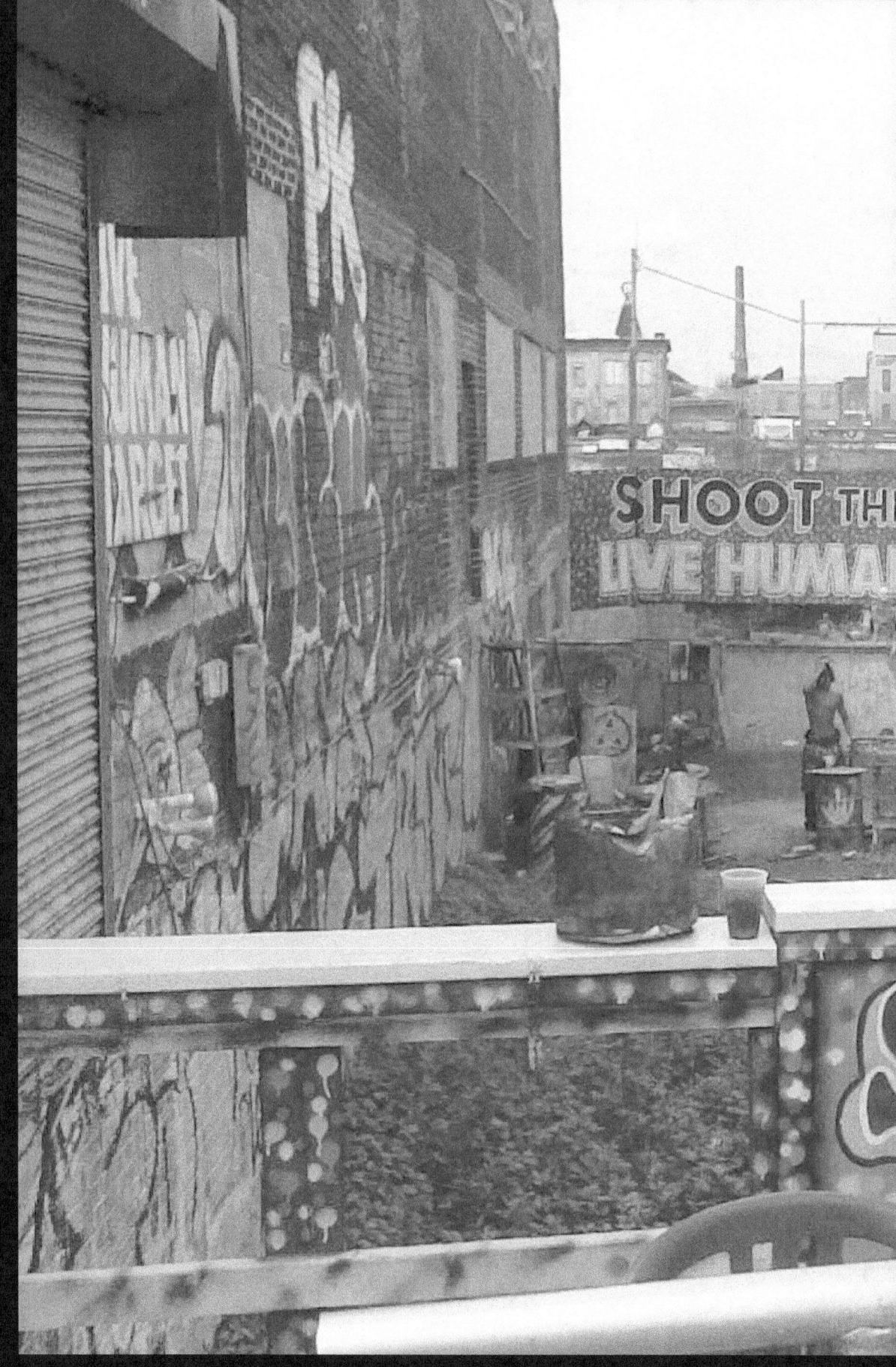

For a while, Shoot the Freak was a weird Coney Island landmark. Located in a vacant lot on the boardwalk, near the now defunct Astroland amusement park, anyone could pay $5 and shoot paintballs at a live human being. Sadly, this metaphor for so much in New York lost its license late in 2010

STYLE WITHOUT SUBVERSION

The V&A's Postmodernism exhibition acted like an industrial trawler, disembedding three decades of cultural artefacts from their diverse ecologies. The result, writes GAIL DAY, *is a deeply conservative reading of this tumultuous epoch*

Most people seem to like the exhibition Postmodernism: Style and Subversion 1970-1990. Such, at least, was the evidence on the night I attended (and from much of the general chit-chat one picks up, and from the cheery presence of promotional leaflets spotted in fashion outlets). Admittedly, I went to the museum on a Friday late night opening when, in the V&A's foyer, a sound system was pumping out music, cocktails were flowing and families were learning a technique for darning holes in jumpers. It was also the night when Charles Jencks was talking to Rem Koolhaas; both have work displayed in the show, so I imagine they must have been in good spirits – and the crowd spilling out from the discussion into the Postmodernism exhibition was generally enjoying the fun of it all.

As may be inferred from my tone, I didn't. Sure, I found plenty of pleasures to revel in – vicarious and otherwise. Tapping toes to Talking Heads, snippets from *Blade Runner* and *The Last of England*, issues of *The Face*, a Buzzcocks' single, and reminders of the Hacienda: it was a retro fairground of an earlier life. Lots of stuff I'd thrown away. My own petty possessions and experiences of the '80s were raised to a second power under the museological gaze named 'postmodernism'. At least I had enjoyed using the commodities back then; with their fetish nature transmuted, they looked back at me from their cultic vitrines and display monitors. Interestingly, the temporal economies invoked by the items of popular culture (the mags, the films, the sounds, the looks) didn't accord with those of the furniture and household objects. If coming across the former felt like rummaging at a jumble sale, the latter was more like window shopping in one of today's emporia,

with their Alessi franchises, devoted to designer products. Not all commodities are equal. Of course, for anyone of my generation, the show inevitably had a melancholic underpinning. But, irrespective of when we were born, Postmodernism treated the reminder of death as a deliberate leitmotif. Jencks' words, stencilled on the wall, set the scene from the outset: if modernism is dead, 'we might as well enjoy picking over the corpse'. Later, Derek Jarman's voice-over was used to echo the sense of historical caesura and closure: 'Even our protests were hopeless'.

The V&A's institutional form shaped the exhibition in two ways. First, the architectural spaces used for the museum's temporary displays forced a tripartite division and, secondly, the focus of the museum's collections gave direction to the type of materials used to typify postmodernism (jewellery, furniture, etc.) The first section largely focused on architecture, drawing on the texts written by architects and theorists who were considered to have initiated the visual and material dimensions of postmodernism: Jencks, Robert Venturi and Denise Scott Brown, and James Wines. (This was not the place to worry ourselves much over Jean-François Lyotard's, Frederic Jameson's or David Harvey's accounts, never mind the arguments of postmodernism's critics.) The second part was largely geared towards a range of design media (furniture, graphic design, etc.). The last section attempted to situate postmodernism in relation to money and the commodity and included, *inter alia*, jewellery, craft and examples of a peculiar high-end phenomenon where architects would be commissioned to conceive a 'piazza' of coffee accoutrements. This final section abandoned the lightbox signboards of the two earlier rooms (red and green respectively) in favour of an abundant use of shiny-black acrylic sheets. The coloured glow associated with commercial promotions in streets and subways was displaced by reflections that conveyed the air of an aspiring celebrity funeral. The exhibition's parts narrated, loosely, the three-fold time frames of postmodernism: its coming to ascendancy,

---

My own petty possessions and experiences of the '80s were raised to a second power

---

its high period and its collapse ('under the weight of its own success'). As a heuristic device, this seemed remarkably conventional. Methodologically, it was something of a dinosaur, especially with the treatment of the final phase as one of internalised self-regard (remember those accounts of the renaissance giving way to mannerism, the baroque to the rococo... or, for that matter, modernism to the international movement?). This conceptual conservatism also emerged via the show's subtitle. Accompanied by its subset of associated binaries (theatrical/theoretical, commercial/ avant-garde, etc.), 'Style and Subversion' was posed as the overarching 'ambiguity' – the all round refusal to be categorised – that was (allegedly) postmodernism. Postmodernism, we were told, was 'a new self-awareness about style itself'. But it transpired that Postmodernism, the show, reduced 'style' to an unreflexive, art historical category which was used to pin down a period of 20 years: strange to see because, if the debates over postmodernity did one thing, it was to distinguish 'ism' from 'ity'.

One would be hard pressed to know

from Postmodernism that the period under scrutiny saw a massive assault on working class communities and labour organisations; significant battles over racist and sexist discrimination, gay rights, abortion rights and anti-Nazi activism; the deregulation of the financial markets; the beginnings (in the UK) of the attacks on free university education and the dismantling of the postwar welfarist settlement. The categorial blanketing performed by 'postmodernism' evaded the specificity of the objects *qua* material objects, let alone the objects as socially situated entities or actors. To reprise the earlier conjuncture: Jarman's angry lament, eight years into Thatcher's term, was fundamentally at odds with Jencks' suave ease and intellectual game playing. It is critically lazy to dub such differences 'ambivalence'. Postmodernism-as-idea effectively bludgeoned into subjection every object presented. There was insufficient recognition that while some claimed to *be* postmodern, self-identifying as its promoters, others only became identified as 'postmodern' by dint of being turned into the tokens within the arguments of the time. For a good number of the exhibits, the label 'postmodern' seemed to be being freshly applied; just being a product of the '70s or '80s seemed sufficient. That surrealist inspired Buzzcocks' cover? It was news to find that the youth of the '80s 'experienced postmodernism for the first time through issues of *The Face* and *i-D*'. No! The curators' opening statement ducked the point: 'This era defies definition, but it is a perfect subject for an exhibition.' Which era doesn't 'defy definition'? Clearly, we were meant to answer 'modernism'. Empty truisms of this sort peppered the show – along with their associated straw target – while reheated paraphrases ('we are all postmodern now') fell short of carrying off pastiche.

Bricolage was a running theme. Loraine Leeson and Peter Dunn's *Big Money is Moving In*, from the project 'The Changing Picture of Docklands' - an intervention in the radical ('left-modernist') montage tradition - found itself reduced to an example of 'postmodern technique'. Working with tenants' action groups and trade unions to oppose the gentrification and corporatisation of their neighbourhoods, the artists developed a series of large publicly sited photo-murals. Shorn of its resistive voice, its activist identification and its commitment to collective agency, the work just served to underscore the closing section's 'money' theme. Earlier we were told that, while modernism created unified wholes, postmodern montage was variable, apparently 'embracing the full diversity of the world'. The 'modernists' named as exemplary of the 'synthesised' mode were Hannah Hoch and Kurt Schwitters. Had the curators actually looked at their work? Along the same lines, we learnt that modernism equates to the grid: a bizarre statement to make when you include, as an example of postmodernism, a celebratory riff on Manhattan's street pattern (Koolhaas' *Delirious New York*). The force of the 'postmodern' chopper came down, conceptually cutting history into trite isms and categories. But - despite what the literature promoting postmodernism claimed - the history of montage practices does not divide itself up into an era of unities and an era of fragments; and neither does the 20th century.

Nevertheless, it was surprisingly interesting to encounter key tokens from its discourse. Despite serving as postmodernism's ideological juggernaut (or perhaps because of it), the architectural material proved especially fascinating. Even the reconstruction of Jencks' Garagia Rotunda, his automotive 'garden folly' dressed up in stagey classicist motifs - though it did strike me all the more powerfully as boringly naff - provided a welcome opportunity to see that confirmed 'in the flesh'. Hans Hollein's line-up of Doric columns (his Strada Novissima, originally made for Venice's Architecture Bienniale in 1980) left me similarly unmoved, if still glad to have registered it as a material presence. The clever ironies now all look so earnest, overweening and portentous. The claim of postmodern architecture was to recover 'meaning' by using historical references, but this populist move just revealed the vacuity of the gesture - its emptiness both *as* gesture and as historical intervention. As lived experience, so-called postmodern space is scarcely different from that of the buildings and squares it sought to trump. However, the discourse on postmodernity advanced itself not so much through its constructed realities as it did through architectural photography. Images in glossy journals like *AD* endowed the examples of 'postmodern architecture' with optical allure and faux clarity. The exhibition didn't give us these, but quite a few architectural models, which captivated in another way, returning us

---

Not all commodities are equal

to the moment of imaginative projection; to the particular totalising perspective of corporate clients and to the encounter in which the architect attempted to convince them that one grand vision could coincide with another.

Nils-Ole Lund's *The Future of Architecture* (1979) was similarly intriguing. Its collagistic fracture was invariably ironed out by the mass circulation of the printed page and then hyped up further and projected by its circulation as a lecture slide. Both reproductions rendered the

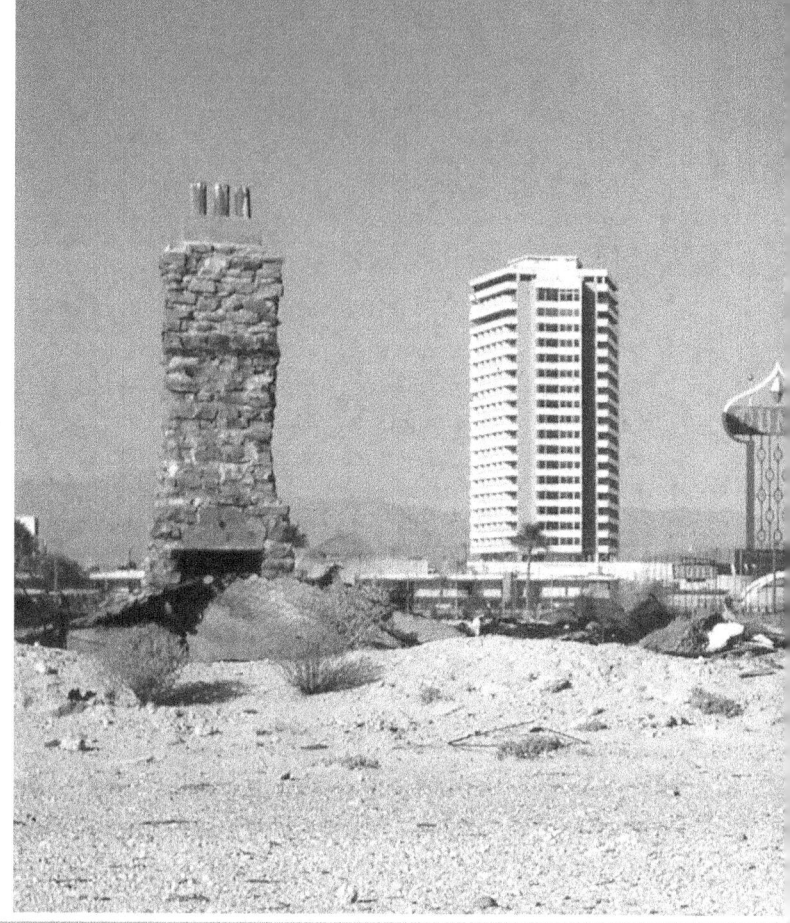

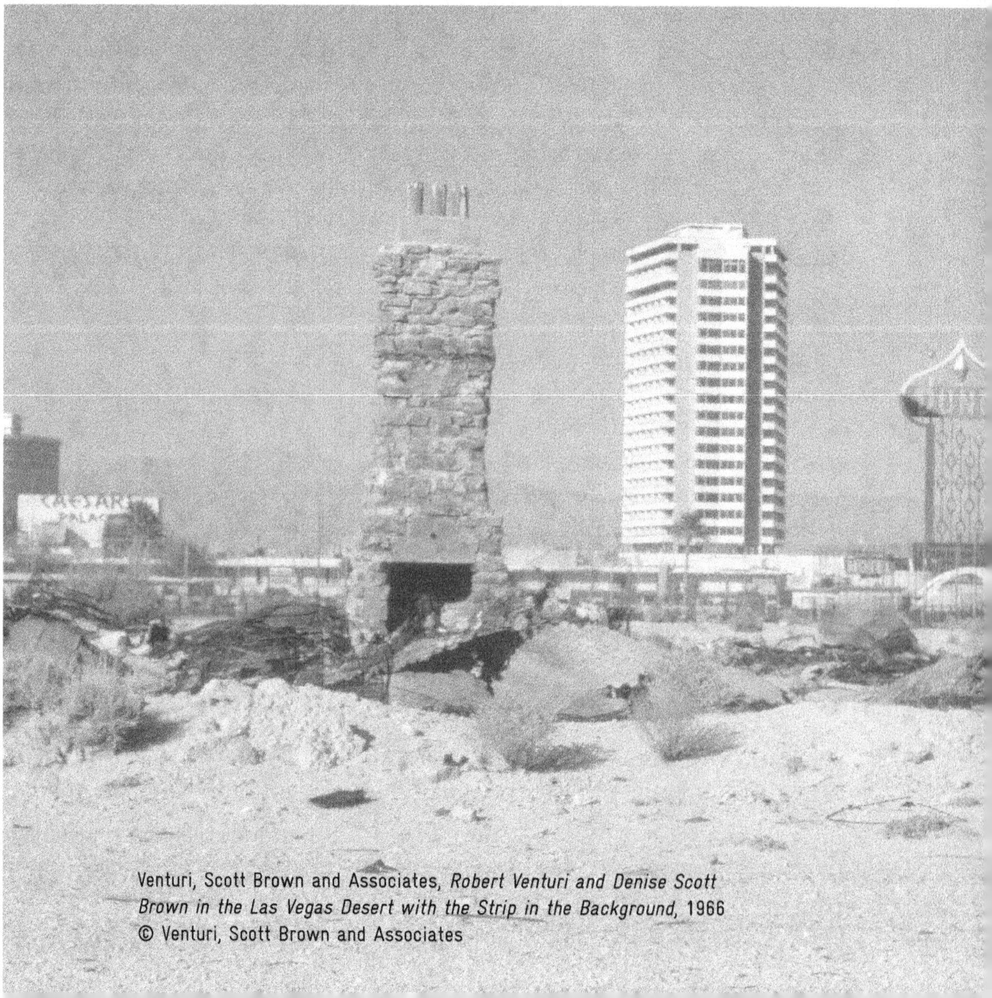

Venturi, Scott Brown and Associates, *Robert Venturi and Denise Scott
Brown in the Las Vegas Desert with the Strip in the Background*, 1966
© Venturi, Scott Brown and Associates

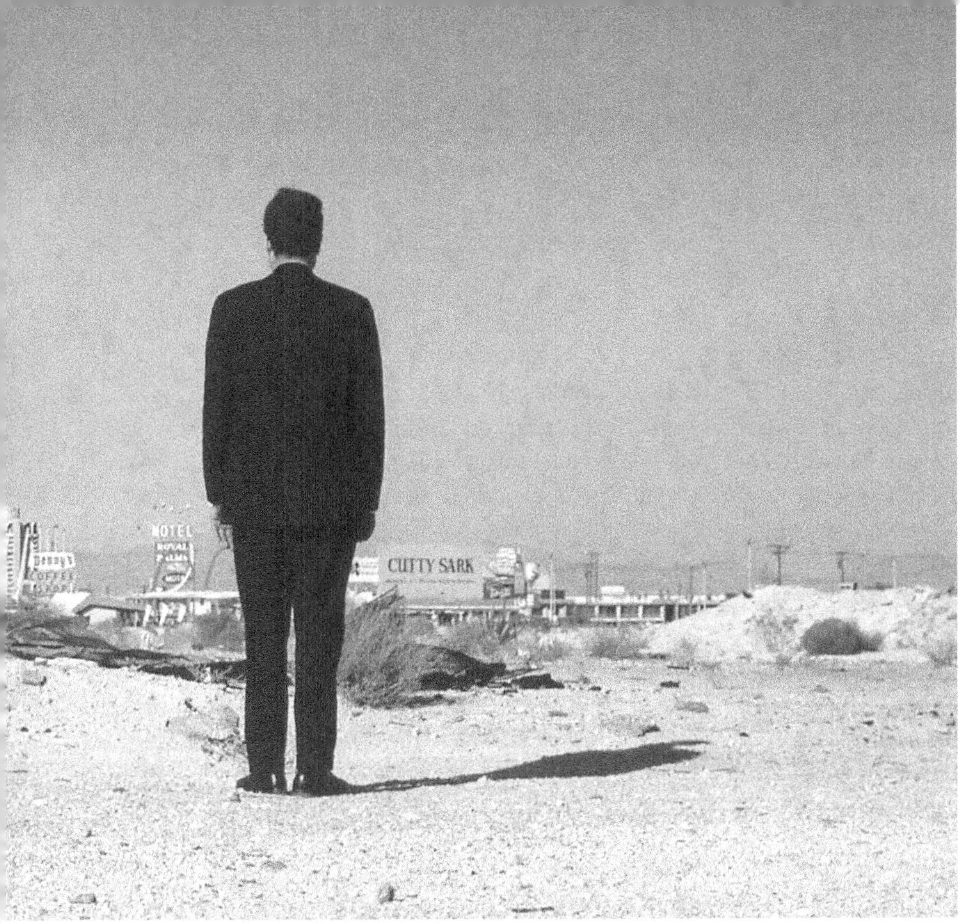

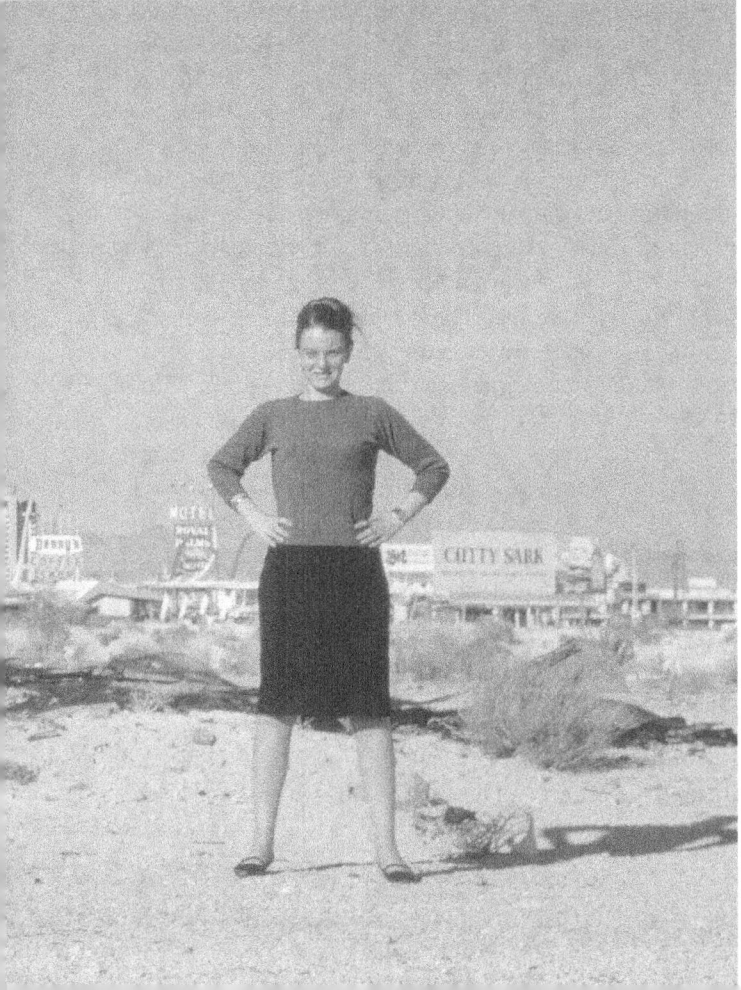

work into something that looked like a large photorealist painting, albeit one with some uncanny transitions on its synthesised surface. Seeing the original montage, I was struck by just how small it was, but I was mostly caught up wondering how it had even become a coin of postmodern discourse. It's so clearly possible to have it read otherwise. Its figure of 'modernity in ruins' could just as easily be interpreted as picturesque, romantic or surrealist; comprehended not as modernism's 'death', but as modernity itself. (These days, we might note, the figure is much favoured by polemicising pro-modernists.)

I came to realise that the cadaver being scavenged was not modernism's but that of postmodernism itself. Perhaps Jencks' opening statement approached the status of a larger curatorial conceit? In one of the show's central rites of passage – its 'Times Square' – the exhibition design echoed the mediatised city of *Blade Runner*; we were taken into that night time world of screens and monitors conjured up so memorably by Ridley Scott, but here it

---

Postmodernism-as-idea effectively bludgeoned into subjection every object presented

---

was David Byrne and Annie Lennox who were the ghastly visages bearing down and cynically mocking us. We had only just passed the *Blade Runner* clips which, in turn, had been placed under remit of 'Apocalypse then'; all suitably 'intertextual'. It would be nice to acknowledge such thematic introjection as a piece of curatorial élan; a meta-joke where an object of study became the exhibition's conceptual

motor. Sadly, the pattern of auto-ingestion was too weakly drawn. Instead, what prevailed was a banal subject/object collapse that seemed not to be of anyone's choosing or staging, but rather the result of a simple failure to maintain critical distinctions or to exercise historical caution.

Certainly, the role of publications (*Domus* or *i-D*, etc.) as disseminators of styles was grasped. But there was also too much uncritical acceptance of what had been read in the canonical books promoting postmodernism. Yet the making of 'postmodernism' – via this avalanche of secondary literature – is a history in its own right. It was a publishing category tied to the sale of pedagogic shorthands. (In the '90s, it used to be said – apocryphally, no doubt – that if you wanted to get your book out, you needed the P-word in the title.) There were allusions to *The Language of Postmodern Architecture* and *Complexity and Contradiction in Architecture* (primary texts, we might say, and which played their part in the construction of the discourse), but the exhibition's perspective seemed to have been essentially directed by the mass of generic style surveys produced to support the art and design school curriculum. These synthetic works were reliant on weakly conceived Wölfflin-type schematisations, derived from Venturi and Jencks and shored up by snippets from the heavyweights like Lyotard or Jameson. Studying 3D design? Well here's how the story goes: Sottsass, Memphis, Studio Alchymia, etc. – irony, pastiche, bricolage, appropriation, quotation. Each segment of the university and polytechnic divisions of labour had its corresponding volumes to enforce and shore up these schema. For all the talk in Postmodernism of deconstruction and reconstruction, of irony and self-awareness, there was little sign that the archive itself had been recognised as form, as institution, as construct or as discursive production – let alone historicised or subjected to critical analysis. The annals were read straight, as direct access to an authentic voice, and then played back.

And yet even the archive was strangely limited. What, for example, happened to the central discussion of the '80s: the distinction between conservative historicist and critical poststructuralist versions of postmodernism? It is rendered by the curators as postmodernism's 'theatrical' and 'theoretical' characteristics, its exciting 'ambivalence'. Yet, at the time, the differences were articulated as sharp political contestations. From Hal Foster's essays in the '80s to the architectural debates of the Revisions group in New York, this sense of embattled opposition was expressed explicitly and repeatedly. We can now recognise these as responses to the Reagan administration's aggressive imposition of neoliberalism. I happen to think there were problems with these left leaning challenges to postmodernism's mainstream, but they were certainly not just one inflection within the postmodern ambiguity.

Less visible, but implicit in the dates and places – and sometimes in the objects themselves – other stories seemed to lurk; further paths suggested themselves for historical investigation. One friend speculated on the possibility that the Milanese design objects – mostly dated immediately after the repression of the Italian left (the so-called 'years of lead') – were symptoms of a 'Pastel Thermidor', Italy's counter-revolution exported and niche marketed. It is tempting to see the pots and kettles as direct symptoms of a new political domination – especially when, like me, you actively don't like them, and especially when you know whose needs they were designed for – but I suspect their place may well be more complex and contradictory. One would want to consider these human products not merely as semiotic representatives of some idiot 'ism', but as material agents within a changing field of social and economic relations. Postmodernism's veneer of historical analysis relied, however, on media coined buzzwords and soundbites: 'yuppy', 'designer decade', 'boom'. History deserves better treatment; so do (at least some of) the objects. The commodities still have their stories to tell. Give the fetishes their due.

Gail Day <G.A.Day@leeds.ac.uk> teaches in the School of Fine Art, History of Art and Cultural Studies at the University of Leeds. She is author of *Dialectical Passions: Negation in Postwar Art Theory* (Columbia University Press, 2010), shortlisted for the 2011 Isaac and Tamara Deutscher Memorial Prize

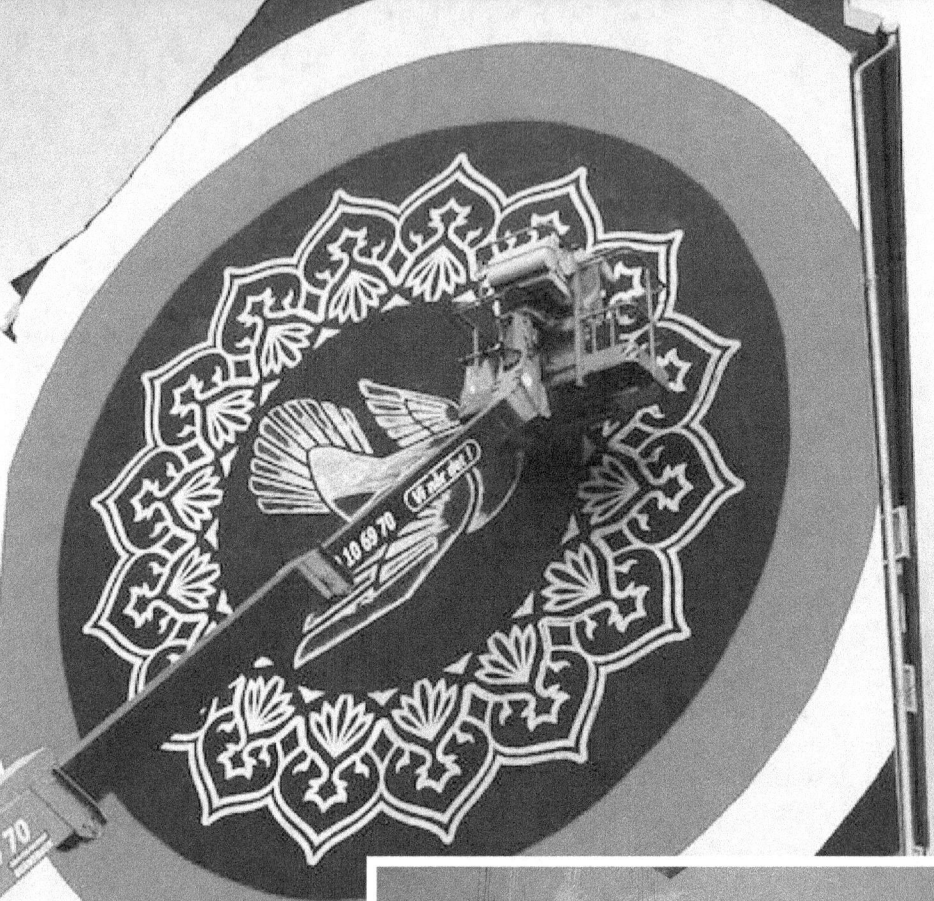

PEACE
69

Like so many other cities, Copenhagen has been subject to the logic of capital through an immense campaign of privatisation, gentrification and normalisation. The symbolic epicentre of these conflicts has been the empty lot where the Youth House, an autonomous squat, stood prior to its eviction and demolition in March 2007

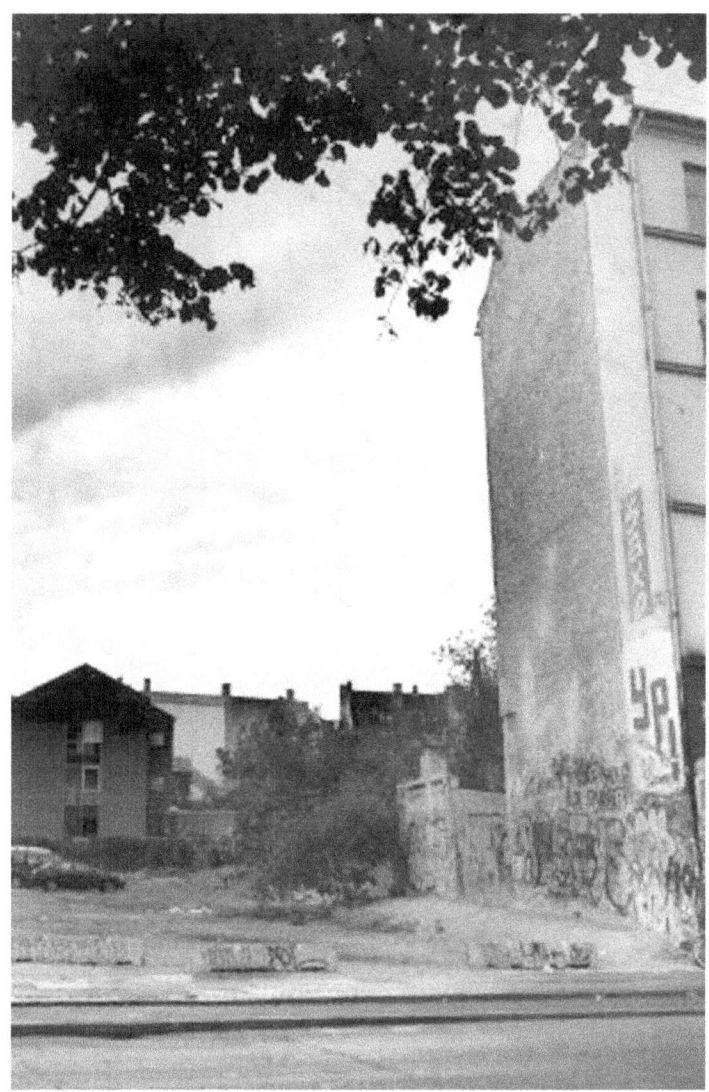

In July 2011, the Copenhagen Municipality commissioned a commercial street artist (no name mentioned) to paint the wall facing the lot, in an attempt to rebrand the site and cover up another injustice. On the night of its completion, the painting was defaced and days later the artist was beaten up. Both the destruction of the mural and the act of violence was widely condemned. But how can direct action take form under the regime of capital? From an aesthetic point of view, one must admit that there is a certain resemblance between a mural destroyed by paint bombs, a battered eye on a hipster and a broken storefront window

REFLECTIONS ON THE ARAB SPRING

Twittering teens or absolutist ayatollahs, men we can do business with or loony autocrats? The media's proliferation of polarities is a strategy to fragment the connectedness of events and disavow western Realpolitik. Here, ANUSTUP BASU reveals the transnational composition of a Spring that is now a Winter

In a column published on 25 May, 2011, *The New York Times* columnist Thomas Friedman issued a pious call to Palestinians. In the wake of the Arab Spring, he invited them to learn from the Egyptian insurrection and adopt the 'Tahrir Square Alternative' (TSA). That is, to announce every Friday a 'Peace Day' and march, in thousands, to Jerusalem, holding an olive branch and a plea for Palestinian statehood, written in Arabic as well as Hebrew, just to avoid any tragic misunderstanding. Implicit in Friedman's conscientious liberalism is a desire for a game-changing symbolic event, one that would insert itself into the sea of information about the uprisings and bring to the fore the image of the peace loving Palestinian finally cleansed of the stigma of pathological fundamentalism attributed to formations like Hamas. The TSA would thus be a transformational strategy that would not just win the hearts and minds of Israel and the world at large but also 'surprise' Benjamin Netanyahu, who, in Friedman's self-admittedly 'crazy' universe, sits in some future anterior moment, reading the column and laughing with characteristic cynicism: 'The Palestinians will never do that. They could never get Hamas to adopt nonviolence. It's not who the Palestinians are.'[1] Friedman of course did not clarify whether the 'surprise', for 'Bibi' would be a pleasant or an unpleasant one.

Secondly, Friedman, in his ardent paternalism, assumes or gratuitously pretends that said strategy has not been already thought of and tried by the Palestinians themselves, including those in Gaza who currently reside in what could well be the largest open air prison in human history. Peter Hart, writing

for FAIR (Fairness and Accuracy in Reporting), has pointed out to Friedman that as a matter of fact Palestinians have, and have for a long time, relentlessly practised the non-violent option without managing to 'surprise' Netanyahu. Such efforts have largely been met with swift and uncompromising repression; they have been responded to with tough love, in the form of arrests and detentions, stun grenades, tear gas, and gunfire.[2] As a 2005 study by Patrick O'Connor established, Palestinian non-violent movements have been overwhelmingly ignored by the free press of the western world.[3] The onus is thus perpetually on the Palestinians, no matter what they do, to emerge – with agon, sacrifice, and endurance beyond human finitudes – as a people capable of some form of newsworthiness that has nothing to do with suicide bombings or crude Qassam rockets.[4] Friedman's invitation towards peace and non-violence emerges from a powerful theme of mainstream American and Zionist common sense that is already weaponised: that Palestinians specifically, and Arabs in general, by virtue of their existence, pose an existentialist threat to Israel and that all their actions, hitherto, have been met by Israel with the solemn purpose of defending itself, no matter what the undertaker says.

The current scenario in Gaza is one of a grotesque human catastrophe perpetrated in slow motion.[5] It is the outcome of a half a decade long vice like embargo and mayhems like the IDF's Operation Cast Lead that, between 27 December 2008t and 21 January 2009, left about 1,400 Palestinians dead and countless injured. And yet, it is this beleaguered 1.5 million strong slice of humanity – plagued by crippling poverty, disease, toxic water, shortage of food and medicine, absence of basic infrastructure and acute unemployment – that seems relentlessly to threaten not just border security, but the very existence of Israel itself. There can therefore be no authentic freedom or exercise of democracy for them without that either bolstering American-Israeli interests or keeping the existent status-quo.[6] As a matter of fact, it seems that there can actually be no recognisable 'people' or territorial notion of 'home' unless these conditions are met. The Death Laboratory of Gaza was a geo-political creation of Israel itself, when Ariel Sharon 'disengaged', removing Israeli citizens and settlements from that space in 2005. With it left the last vestiges of imperially recognised 'peopleness' from that space, one that could be attached to humane concerns about peace, neighbourliness or hospitality. It was that strategic withdrawal that created the possibility of reinventing Gaza as a pure ground zero of 'bare life', as Georgio Agamben would say, where, following a long standing Zionist theme articulated by Golda Meir and many others, the Palestinian people (or for that matter any people) do not exist. What exists is a pathological biomass, an absolute spectre of Islamic terror that needs to be defended against, with the old, infirm and infantile to be dubbed 'human shields'. It is this weaponised and mediatised defensive redoubt that holds paramount status, especially when it comes to territories illegally occupied by Israel from the West Bank to the Golan Heights.[7]

THE SPRING OF THE PRESENT AND THE LONG HOT SUMMER OTHERWISE

I have begun this essay with this grotesque picture from the recent past for three primary reasons. The first one should be fairly evident – that American mainstream media responses (which I will largely focus on) to events like the Arab Spring are guided by a curious mixture of an almost onto-theological commitment to abstract, totemic ideals like 'freedom' and 'democracy' and a *Realpolitik* one to American strategic, monetarist and security interests in the Middle East. In the overall flow of informatised commonsense, the two lines of reckoning are rendered inseparable. Democratic structures of representation anywhere in the world cannot be disruptive in relation to networks of governance and financialisation stipulated by the Washington Consensus. Democracy must yield 'liberalism' in its neo-incarnation; it cannot give us Hamas or the Muslim Brotherhood, instead. My second reason is that the 'Arab Spring' is still unfolding in front of us with a long rumble. It is always difficult and dangerous to 'read' the present, for any understanding of it is already belated. The present has to be grasped in a manner that is open to the many imaginative and political possibilities – of self-making, sovereignty or antagonism – that it brings. My invocation of the Israeli-Palestinian situation has been intended to illustrate a habit of neoliberal statist thinking that, in the name of security, stability and combatting terror, threatens to kill us, anyway. It is this murderous habit of thinking that imperils, more than anything else, the exhilarating possibilities of the Arab Spring.

The third reason pertains to an obvious paradox: unlike the Egyptian or Iranian youngster who apparently just wants to be an American teenager and tweet in peace (much like the American who waited to jump out of every Gook in Vietnam, according to the emphatic Colonel in Kubrick's *Full Metal Jacket*), non-violent *democratic* activists in Gaza somehow have not been able to twitter themselves into the spotlight. Curiously, neither have democratic activists brutalised and jailed in countries occupied by the United States or its allies: Yemen, Bahrain, Saudi Arabia, Sulemaniyah Iraq, Afghanistan or the United Arab Emirates. Media focus on popular outrage expressed on Twitter or Facebook seems to be disproportionately trained on enemies of the West, like Iran, Syria or Libya. It is not that there were simply no tweets from Bahrain when Saudi and UAE forces, armed with their latest military acquisitions from America, marched in to crush the insurrection; it is just that such voices were not deemed 'newsworthy'. That is, they were evaluated as such when 'news' itself in our informational world, as the late Derrida astutely observed, is that in which 'actuality' tends to be 'spontaneously ethnocentric.'[8] It is this instantly consumable, informatic and industrialised ethnocentrism that encompasses not just state policy, but also a media space dominated overwhelmingly by about half a dozen giant conglomerates devoted to global metropolitan interests.

This is not to say that in a world of horizontal connectivities other voices, evocative images of alterity or testimonies of anguish or pain are not registered and shared across the world. The point, however, is that increasingly, in our occasion, statist dominance over the media ecology is exercised not so much in axiomatic, top-down ways through censorship and elimination (although some such efforts exist: George Bush bombed the *Al Jazeera* offices in Afghanistan in 2001; Ben Ali banned YouTube, dailymotion and Takriz; Mubarak shut off the internet and cell phones; China, at one point, prevented recent images of insurrection entering its media space). Instead, dominance is achieved by absorbing errant images and sounds to already-there, massified structures of feeling and perception: orientalism, race,

What exists is a pathological biomass, an absolute spectre of Islamic terror that needs to be defended against

كيفية إستخدام الأدوات

١ - الدرع والدوكو

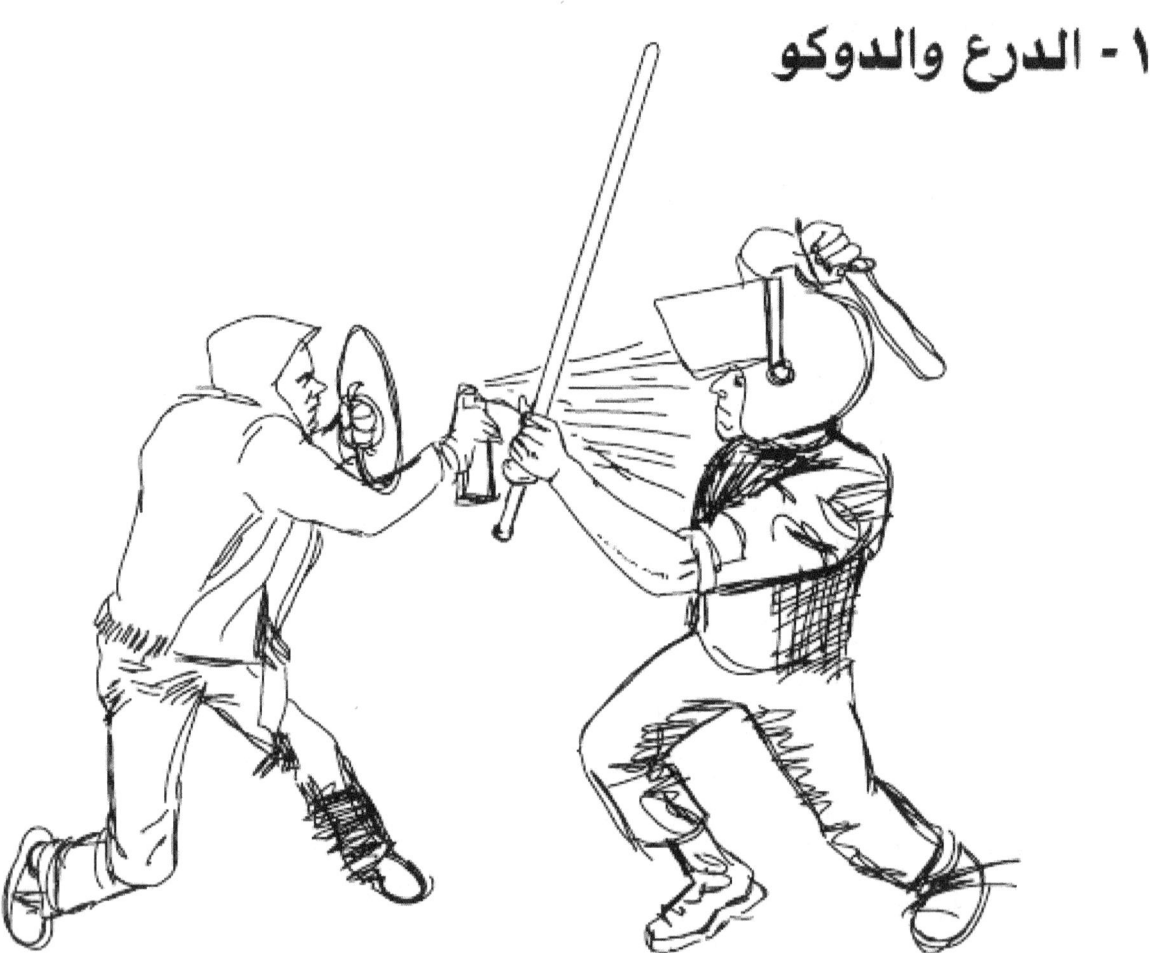

إثبت مكانك يامصري. صد العصاية بالدرع وأنت تقوم بالرش في الوجه.

How to Use the Accessories

Shield and Spray

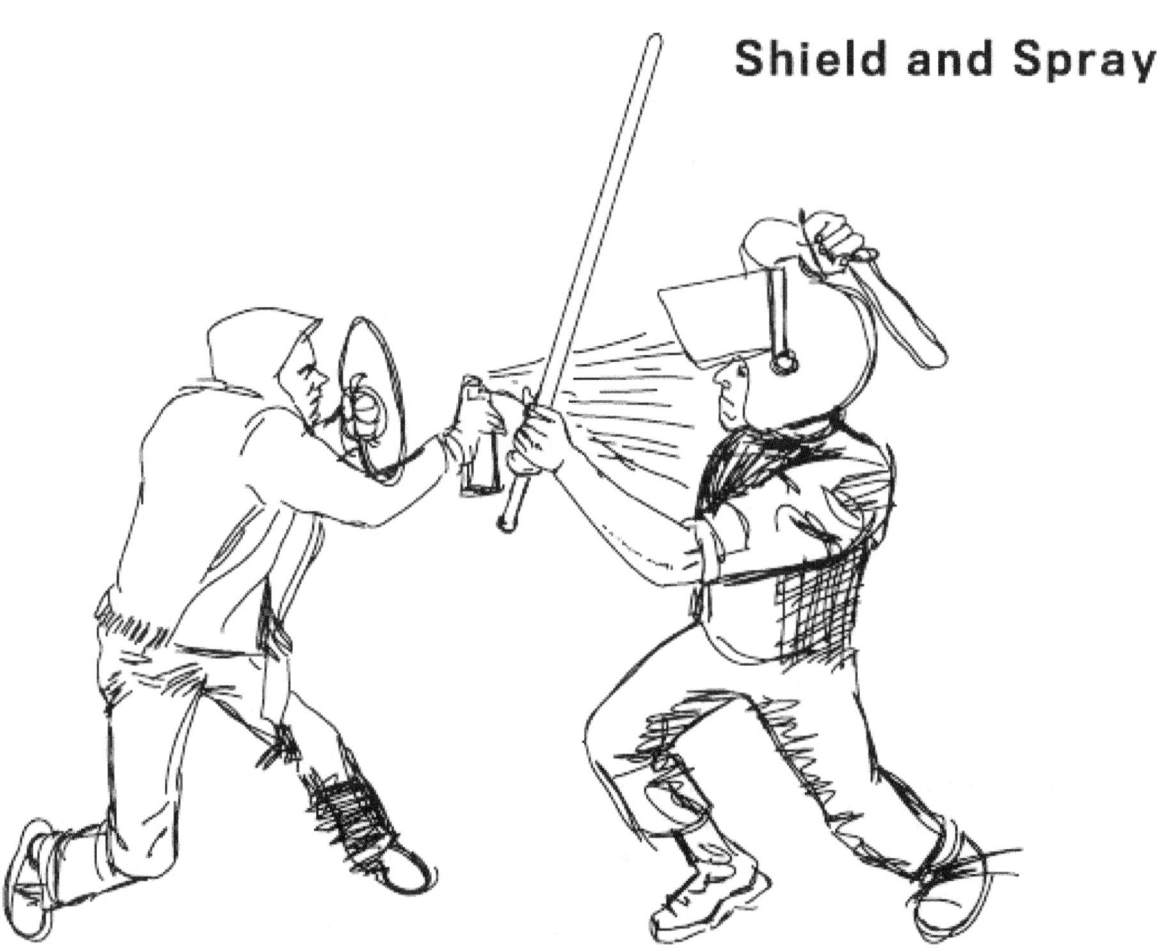

HOLD YOUR GROUND, EGYPTIAN!
Block the truncheon with your shield
as you're spraying them in the face.

terror, security, stability, Islamophobia, pious concerns of Clintonian multiculturalism or anxieties about immigrants. The point, therefore, is not to shut out, in a total manner, images of disturbance, but to absorb them as fresh noises into an overall clamour already enveloped and dominated by axiomatic myths about free market and freedom and about America being the reluctant behemoth of good in a dangerous world. This is how questions of human dignity and liberty become tied to a presiding onto-theology of capital. This is also how relations around Egyptian youth protests are enframed and tempered by the *Realpolitik* fear and loathing of insidious energies hailing from that nebulous thing called the 'Arab Street'. Ergo, it is to be lauded that the former want democracy, but with the latter around, perhaps, that might be too soon.

It is this already techno-deterministic template of information culture that encourages one unquestioningly to distinguish between greater evils and 'practical', 'indispensable' ones, between rhetoric of necessary change and a metropolitan strategic silence in which all of us are invited to be complicit. Bahrain has been ruled by the Sunni Khalifa dynasty for well more than two centuries now. If the world was to turn upside down and the hitherto repressed Shia majority were to gain political prominence, Bahrain, as per a realist-statist world view, would inevitably tilt in the direction of Iran. That would not bode well for the greater cause of freedom in the world since Bahrain houses the US navy's fifth fleet and is strategically important for the control of the Suez Canal and the Straits of Hormuz through which almost a quarter of the oil supply passes. Similarly, despite the fact that President Ali Abdullah Saleh has ruled Yemen for 33 years and has had untrustworthy

flirtations and friendships with Russia, Iran and Saddam's Iraq, his brutal efforts to strike down dissent in his country did not attract as strong condemnations as did Assad's in Syria. While the Obama administration and the Gulf Cooperation Council Bloc has pressurised

---

Western arms industries have consistently supplied these regimes with weaponry not used against foreign threats, but almost exclusively to pacify domestic populations

---

Saleh – presently convalescing in Saudi Arabia after an assassination attempt in early June – to step down and allow the Yemeni people to fulfill their 'aspirations', it remains amply clear that the broader template of regional 'cooperation' cannot allow such aspirations to disturb American military interests in this impoverished Arab country due to its location near the major waterways and Somalia. Saleh has been a faithful soldier in the 'War on Terror'; he has allowed the Obama administration to open – along with Iraq, Afghanistan, Pakistan and Libya – a fifth theatre of conflict in Yemen in which there have been repeated drone attacks to kill the few hundred members which al Qaida in the Arabian Peninsula (AQAP) supposedly has.

In contrast, the 'Arab Spring' has provided a wonderful opportunity for the West to topple Muammar Gaddafi in Libya[*]; until recently this

[*] This article was written in 22 September 2011 before Gaddafi's death

eccentric tyrant must have been envisioned as a sobered up version of his cold war self, for he was deemed trustworthy enough to be granted weapons worth $470 million by the European powers in 2009 alone. For its part, before the calls for change became strident, the US government was working on a weapons deal worth $77 million just to top off the $17 million it provided in 2009 and the $46 million it supplied in 2008.[9] Perhaps too much eccentricity or too much tyranny is not how Gaddafi overplayed his cards. Perhaps he has come to regret the statement he made in January 2009, expressing a desire to nationalise the Libyan oil industry.[10]

Friedman's pious articulation of the TSA, of course, does not take into account either the fact that the United States and North Atlantic powers have been supporting dictatorial or authoritarian regimes in Cameroon (Paul Biya), Turkmenistan (Berdimuhamedow), Equatorial Guinea (Nguema), Chad (Idriss Deby), Uzbekistan (Karimov) or Ethiopia (Zenawi), apart from the usual suspects in the Gulf, or that it has also been providing many of these regimes the arsenal to prevent or exterminate the TSA. Perhaps it slipped Friedman's mind that the Arms Industries of the West (dominated overwhelmingly by the United States) have been consistently supplying these regimes with weaponry that has been used not against foreign threat, but almost totally to keep domestic populations in check. It is the West that has been giving them 'deep packet inspection' technologies through firms like Narus, Ixia or Sandvine to police the airwaves and throttle dissent and subversion. Hence it was not just the 'Made in USA' tear gas canisters used in Tahrir and dolefully pondered over by talking heads in mainstream American media, but also the live ammunition, armoured cars, helicopters and tanks used to crush the TSA in the Pearl Square in Manama, Bahrain. In this case, it was not just the people gathered to protest, but the Square itself that was eliminated.

THE LOGIC OF TELELOCALISATION: WHAT IS SO 'ARAB' ABOUT THE SPRING?

From discourses of governance that abound in print and electronic media, it has become apparent that a new Egyptian dispensation has to prove that it is capable of handling 'freedom' and 'democracy' responsibly by sticking to some essential things: continuing to be a client state of American-Israeli interests, maintaining the Camp David accords and aiding in the blockade of Gaza[11]; keeping the Suez Canal accessible to western powers and closed to Iran; securing the crucial pipeline that supplies natural gas to Israel and other Arab nations. When Mubarak closed the Rafah Crossing more than three years ago to strengthen the deadly embargo on Gaza, it was a deeply unpopular move in Egypt. It is expected that his successor will continue to do such things no matter what the 'people' say. The 'revolution' is thus expected to shrink and step back into an already awaiting straitjacket of 'responsible reform', one that will keep certain planetary structures of financialisation and war in place. It is, therefore, 'telelocalised' from the onset as a local rumble that must eventually be diagnosed, bracketed off and absorbed into the great administration of things. The Egyptians, for instance, could warily look southwards and recall the grotesque overwriting of Nelson Mandela and the African National Congress' Freedom Charter by powerful western financial institutions after the end of apartheid.[12] They could also remind themselves that fresh updated versions of the neoliberal 'shock doctrines' are usually tried out first in the peripheries

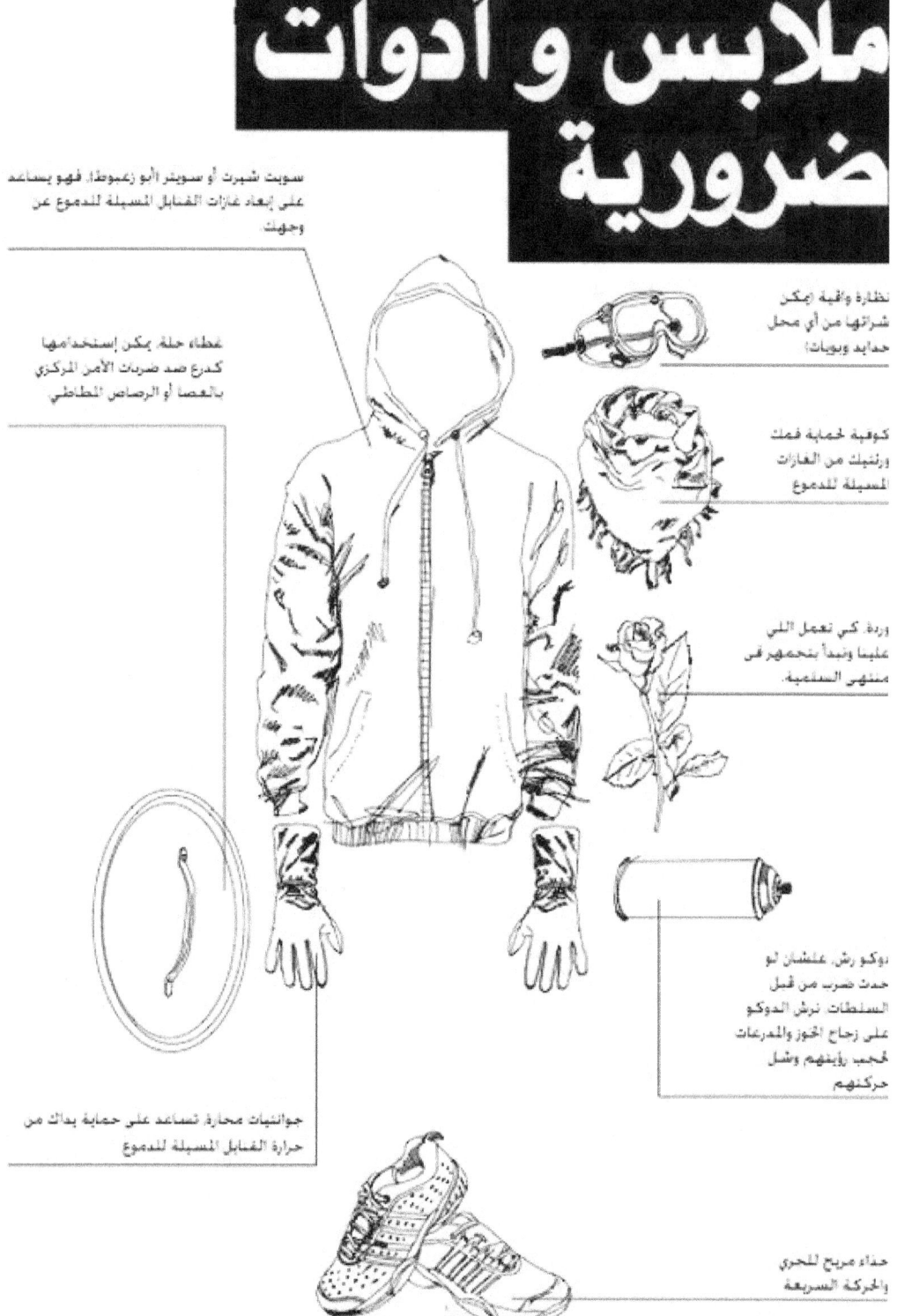

ملابس و أدوات ضرورية

سويت شيرت أو سويتر أبو زعبوطة. فهو يساعد على إبعاد غازات القنابل المسيلة للدموع عن وجهك

نظارة واقية يمكن شراؤها من أي محل حدايد وبويات!

غطاء حلة يمكن إستخدامها كدرع ضد ضربات الأمن المركزي بالعصا أو الرصاص المطاطي

كوفية لحماية فمك ورئتيك من الغازات المسيلة للدموع

وردة، كي تعمل اللي عليها وتمداً بتخمهير في منهي السلمية.

دوكو رش. علشان لو حدث ضرب من قبل السلطات نرش الدوكو على زجاج الخوز والمدرعات لحجب رؤيتهم وشل حركتهم

جوانتيات محارة تساعد على حماية يداك من حرارة القنابل المسيلة للدموع

حذاء مريح للجري والحركة السريعة

Necessary Clothing and Accessories

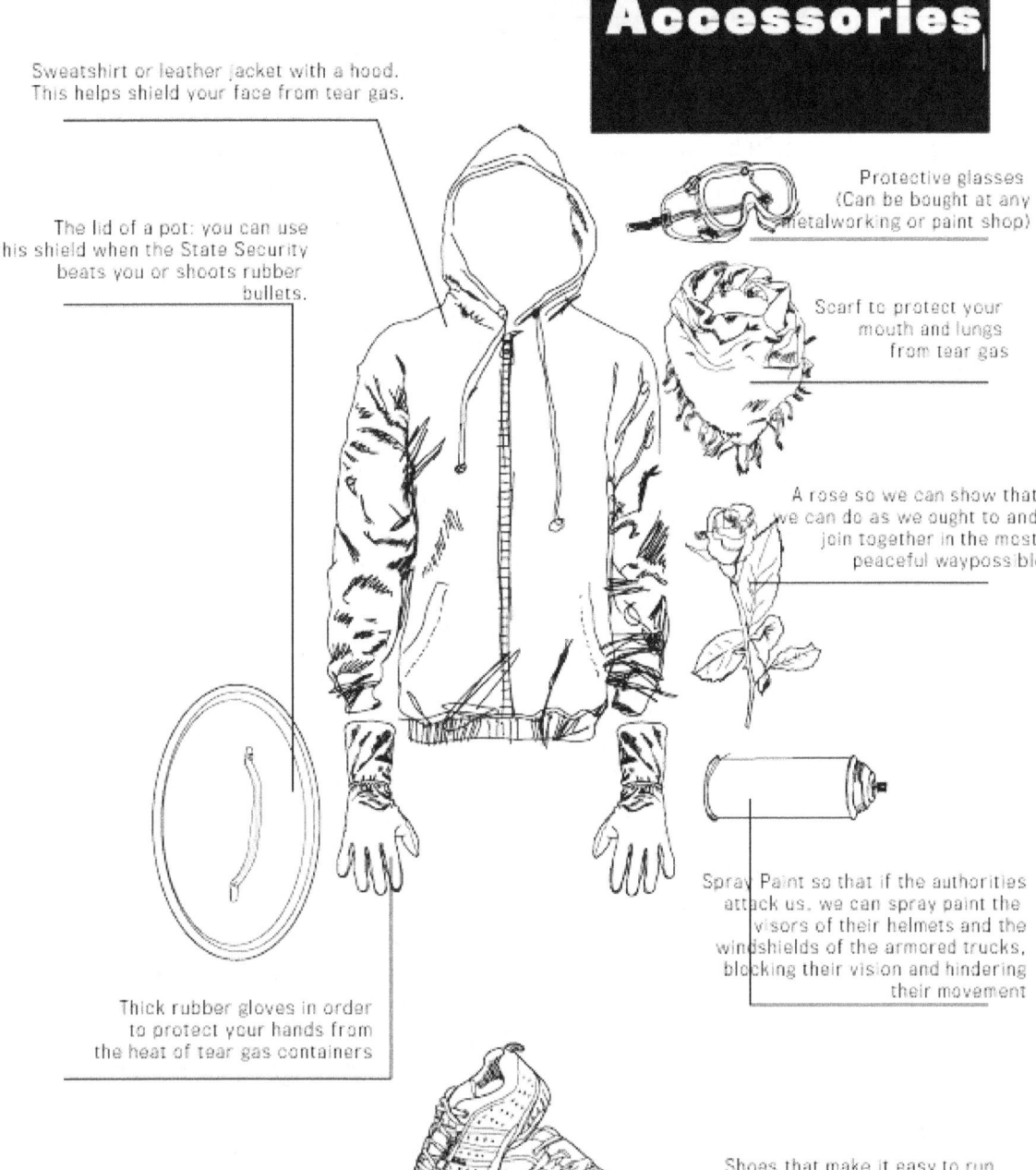

Sweatshirt or leather jacket with a hood. This helps shield your face from tear gas.

The lid of a pot: you can use this shield when the State Security beats you or shoots rubber bullets.

Thick rubber gloves in order to protect your hands from the heat of tear gas containers

Protective glasses (Can be bought at any metalworking or paint shop)

Scarf to protect your mouth and lungs from tear gas

A rose so we can show that we can do as we ought to and join together in the most peaceful waypossible

Spray Paint so that if the authorities attack us, we can spray paint the visors of their helmets and the windshields of the armored trucks, blocking their vision and hindering their movement

Shoes that make it easy to run and move quickly.

We have long thought the revolution will be televised; we've only recently started wondering what it means for a revolution to be 'informatised'

rather than in the metropolitan centres. Statist neoliberalism, as a matter of fact, was first tried out in Pinochet's Chile after the coup d'état in 1973, more than half a decade before it became the template in Thatcher's England or Reagan's America.[13] There are strong indications that something similar is presently being attempted in Iraq.

As per a panoptic point of view of neoliberal governance, all forms of self-making, desire and hurly-burly of protest must finally yield to the civic religiosity of North Atlantic market structures. Ronald Judy has identified this planetary form of sovereignty as that which is 'the realization of perpetual change and a pre-emption of change at the same time.'[14] The only firmament of transformation that is thereby allowed is that of the 'free market'. Apart from American-Israeli geopolitical interests (and their bywords for these like 'security', 'stability', etc.): 'responsible freedom' also means following some already awaiting imperatives of military-industrial finance: the proper handling of the annual three billion dollar US military aid to Egypt and continued issuance of lucrative arms contracts to Lockheed-Martin, Boeing, General Dynamics or Raytheon.

It was in this spirit that the western powers, encouraged by Mossad, first called upon Hosni Mubarak himself to be the midwife of 'change' and, failing that, attempted to put Omar Suleiman in his place. Having headed the Egyptian General Intelligence Service (EGIS) since 1993, Suleiman was not just instrumental in choking dissent among his own countrymen, but also the chief supervisor of 'extraordinary rendition' programmes that the CIA delegated to him.[15] That, too, failed and the scenario, under military control, is still an unfolding one. However, it can safely be said that the Egyptian people can expect many such a tip of the hat.

Quite a few of them, as Karl Marx observed in a different, but exemplary context more than 160 years ago, will be that of the Napoleonic three-cornered one.[16] For the moment, apart from the geo-strategic concerns already mentioned, the West will be watching with avid interest how, in the new dispensation, the Egyptian economy will be structured, given that the entity in power, the Egyptian armed forces – the beneficiary of more than 40 billion dollars from Washington since 1979 – virtually dominates all its sectors.[17] The International Monetary Fund has already granted a loan of $3 billion to the interim government, after consistently praising the elite kleptocracy headed by Mubarak over the years for pushing through neoliberal measures and devastating the Egyptian population.[18] There are also growing concerns about the future of labour rights, press freedom and the rights of women and minorities in the new dispensation of 'stabilisation' and 'modernisation' that is coming into being.

We have long since pondered whether the revolution will be televised; it is only lately we have started wondering about what it means for a revolution to be 'informatised'. The latter is a relatively new architecture of power in our times; it entails a managing of popular energies and worldly humanitarian and political concerns by ascribing a human face to 'change', giving a proper name to Mephistopheles (Ben Ali, Mubarak, Saleh, or Assad) as well as the Messiah (Mohammad ElBaradei or perhaps Google's Wael Ghonim) and then restoring the catastrophic balance of imperial interests. Cranky old patriarchs can have their Autumns; people can have their Springs; iron death masks of power are eminently expendable or changeable beyond a point; but the planetary military-industrial-techno-financial assemblage is not. The power of informatisation seeks to 'telelocalise' a milieu of unrest from an almighty metropolitan perspective; it seeks to invent the 'people' as well as manage, dictate and name its 'aspirations'.[19] This it does by polarising themes (Egypt contra Iran, twittering teens contra absolutist ayatollahs) or collapsing them together (Muslim Brotherhood plus al Qaida plus Taliban); making instant and vulgar comparativist evaluations (a 'secular' tyranny is a lesser evil than Islam/Terror); and curtailing the historical horizons of possibility by drumming transcendent abstractions like 'security,' 'order' and 'stability'.

The social power of informatisation draws its powers from a mythical, cosmic perspective it has claimed for itself. It is from these commanding heights that it 'invents' and represents a 'locale'. It is necessary to 'represent' something because, unless something is represented, it cannot be governed. Consider the statement made by Tony Blair in January, 2011 on BBC Radio 4, distinguishing between Mubarak and Saddam Hussein: he said that the two cannot be called comparable dictators because Mubarak has presided over an 'Egyptian economy' that has doubled in the last decade or so.[20] That factor, along with Mubarak's strong military support for western interests beginning with the first Gulf War, makes Egypt a theatre in which only the logic of economism and that of the 'War on Terror' need apply. In this majestic abstraction of Egypt in relation to world affairs, it becomes a matter of very small print that, officially, more than 22 percent of the population live in abject poverty (less than $2 a day), with an equal number very close to it; that the rate of unemployment is close to 10 percent and more than double that amongst the youth; and that common people, in recent years, have been hit by an inflation in consumer prices that perpetually hovers close to 12 percent.[21] Like in many similar scenarios, these official statistics do not account for the current global malaise of underemployment. Shortly before the eruption of the 'Spring', there were demonstrations in Egypt calling for a monthly minimum wage of 1,200 Egyptian pounds; the kleptocratic government full of businessmen could promise only 400 LE, which amounts to about $67.[22] Blair's sweeping statement, in a figurative sense, comes from the same telelocalising heights of American drones in Yemen or Pakistan, whose operators sit in the Creech Naval Base in Nevada or Langley, Virginia and bomb populations after abstracting pictures of 'terror' through what is known as 'pattern of life analysis.'[23]

Telelocalising a milieu also means to

Telelocalising a milieu also means to provincialise its narrative

provincialise its narrative; to make Egypt's story absolutely its own. It is to enwrap the milieu of unrest into cocoons of national, regional or ethnic scenarios and not extend it to a world swept by uprisings and demonstrations from Mexico, Haiti and Honduras, to Madison and California in the United States, to Spain, Britain, France, Italy, Portugal or Greece in Europe. Why is it that the protests in Cairo or Alexandria have to be categorically isolated from the Tent City movements in Israel, tribal assertions against the government and mining multi-nationals in India, or the thousands who marched along the Des Voux Road in Hong Kong? Why is it that Friedman's TSA can advance bravely along the Arab Street, but the street itself has to end as soon as it approaches absolutist oil rich allies like Saudi Arabia, Bahrain or the UAE? The primary impulse of informational

news is to promote parcelled accounts of such eruptive events based on ethnographic, already existing diagnoses about which societies are mature enough to make an 'orderly transition' from authoritarianism to democracy (Iranians, perhaps Egyptians) and which people are clearly not (the Saudis). It is to engraft them agonistically into a singular unfolding narrative of capital in the world and foreclose a possibility of them merging with stories and histories that are different. The logic of 'information', taken in this special sense, is to present things as 'already shot through with explanation', as Walter Benjamin once said.[24] This it does by nulling historical complexity, abolishing critical memory and reducing language to a set of linguistic functionalisms. When was the last time that viewers of Murdoch's Fox News Channel, or even the more 'liberal' CNN, were reminded that it was only in 1953 that Iran had a democratically elected socialist Prime Minister?

So what is so essentially Arab about the Arab Spring? Why are the rumbles in the Arab world and distant thunders elsewhere symptomatic of not just this or that regime's long pending disintegration, but of the planetary narrative of the Washington Consensus itself coming apart at the seams? Is the story simply a vulgar Freudian psychodrama of hitherto infantile but now slightly mature populations killing inclement fathers or demanding dignity or recognition from them? Why are the recent events in London to be deemed absolutely distant from the 60-odd food riots that, according to the US State Department, took place across the world in the last two years alone? In the month of April, 2011, the World Bank president, Robert Zoellick, said that the global economy is 'one shock away' from a catastrophic crisis in food supplies, estimating that in the last year alone 44 million people had 'fallen into poverty due to rising food prices'.[25] The United Nations' FAO Food Price Index (FFPI) averaged 234 points in June 2011, 1 percent higher than in May and 39 percent higher than in June 2010.[26] It had reached its peak at 238 points in February. This scenario of devastation is as much the outcome of expanding deserts, falling water tables, droughts and famines or increasingly hot Summers as it is of deregulated speculation on commodity futures and oil prices.[27] It stretches from Haiti to Algeria, on to India and across up to the Philippines. It is tragically compounded by the fact, that while the techno-financial elites of Wall Street and its satellite formations across the world have long since recovered their fortunes lost in the downturn of 2007, according to the International Labor Organization's 2010/11 Report, growth in global wages slowed from 2.8 percent in the beginning of the crisis to 1.5 percent in 2008 and 1.6 percent in 2009. If China is taken away from the picture, the figures come further down to 0.8 percent in 2008 and 0.7 percent in 2009.[28]

How indeed can movements in London be insulated from the hot winds blowing in from Cairo? Perhaps, the Egyptians may take heart in the fact that, despite their tribulations, according to the CIA's World Fact Book, they, in being placed 90th, rank much lower than the US (39th) and are almost on a par with Tony Blair's England (92nd) as far as income inequality is concerned.[29]

CONCLUSION

There have to be other forms of reckoning with such world wide eruptions of antagonistic energy and affect. The global landscape of violence has to be mapped in coincidence with an equally expansive map of North Atlantic financial elites, their constable states, satellite plutocracies and techno-managerial oligarchies across the world. This is not just a landscape of gross class exploitation and debt enslavement, but also one in which not just populations, but entire forms of life can be systematically rendered 'disposable' in an instant, by long distance speculations, remotely controlled 'structural adjustments' or, in a more elementary manner, by predator drones. The rice farmers in the Philippines perhaps know that instinctively without tracking the intricacies of tariff walls, as do tribal folks in Central India who have been asked to vacate their habitats and the bauxite rich mountains they have been worshipping as gods for centuries; hungry populations in Ethiopia or Sudan discern that something is rotten when their governments sign surreptitious deals, leasing arable land to distant powers like South Korea or China; the rural people of the Qandahar and Helmand provinces in Afghanistan create their own cosmologies of meaning and affect given the fact that, according to a recent poll conducted by the International Council on Security and Development, 92 percent of males (women were not polled) do not know anything about 9/11 and 40 percent believe that the war it triggered was on Islam, with the rest concluding that it was on Afghanistan.[30]

Arjun Appadurai has talked about a global intuition of poor people.[31] According to him, the cellular, osmotic powers of the financialisation of the planet operate insidiously, often minus the sound and fury of the clear and present nation state and its vertical instruments of welfare and repression. However, perhaps, at an affective-popular level, the processes and outcomes of neoliberal globalisation are being questioned in myriad ways, bringing them into critical proximity with past horrors of colonial genocide, enslavement, exploitation and development of underdevelopment by the rapid devaluation of local modes of production. I call these formations *intense* localisms, keeping in mind the etymological variant *intendere*, which means to intend. Intense localisms therefore, are local cosmologies of justice that emerge from clashes between alien inflictions coming from a distance, and rooted customs, juridical and theological perspectives, stories, world views, solidarities, and affectations.[32] These determinations of justice are, in most cases, *intended*, despite the fact that planetary metropolitan narratives of governance, security and news (the powers behind which flout international law with impunity) attempt to overwrite them as directionless, chaotic, or pathologically beholden to 'terror'. Intense localisms come to the fore in a world in which desire is democratised, but the means to it are acutely monopolised; in which the mobility and bargaining power of labour is brutally restricted, but the movement and reach of capital is extended to the production of social life in and of itself. Intense localisms have thus emerged in an hour of the abject dismantling of the postwar welfare state, the financial subversion of the postcolonial state through comprador elites, withdrawal of social security programs, rampant privatisation of all sectors including health, education and natural resources, the planetary technologisation of agriculture, extermination of the commons, abject formalisations of the very concept of citizenship, and summary destructions of ecological habitats and scenes of nativity. Each of these cosmologies of justice are unique in some way, yet they are also contiguous to each other. In their separate ways, they have called the new world order to judgement.

This planetary swell of antagonistic energies undoubtedly takes both good and bad

forms. Some of them merge with movements attempting to forge a politics of the new (from vital student mobilisations in Chile to the *Indignados* in Spain, to youth agitators in Tel Aviv or Hong Kong); others are captured by state machines. Examples of the latter would be the folksy righteousness and suicidal statism of the recently minted American Tea Party, the swarm of ultra right, racist and anti-immigrant populisms in Europe (the BNP and the English Defence League in Britain, Le Pen's Front National in France, Geert Wilders and the Party for Freedom in Holland, the Jobbik in Hungary, Jörg Haider in Austria, or the best-selling neo-fascism of Thilo Sarrazin in Germany); or even, in a different sense, the current undemocratic agitation in India led by Anna Hazare that calls for combatting corruption by the setting up of an absolutist extra-judicial paternalistic authority called the Lokpal.

Similarly, it must be said that good and bad impulses of a many armed, billion strong Islamic faith in the world will indeed intensely shape and influence such local cosmologies and moral economies. How can one justify cynical calls to control transformational possibilities in Egypt or Gaza because of the spectral Muslim Brotherhood or Hamas (which also happens to be the largest and most efficient humanitarian organisation in Palestine) when the American and Israeli democracies continue to be strongly impelled by hard right Christian groups and Zionist parties like Likud? Affirming the historical and political valence of Tahrir Square is to grasp it without any existing assurances about 'security' and 'stability' and ready at hand fears about Iran; it is to be critically open to its possibilities, both good and bad. It is, as Alain Badiou recently reminded us, also to approach it like a student and not some stupid pontificating professor, precisely in order to freshly learn the very ways of distinguishing the good from the bad.[33] All parties in Tahrir Square – the students, the Marxists, the Nasserites, the incredibly brave women, and indeed the Muslim Brotherhood – have been and will continue to make history. And yet, as Marx observed in relation to a different scenario in the past, perhaps none of them will make it of their 'own free will; not under circumstances they themselves have chosen.'[34] Some such efforts will be tragic, some farcical and some victorious, but if there is indeed hope in the Arab Spring, it is that there will be collective energies that will keep renewing themselves and returning to break the dead calm of things.

Making a distinction between good and bad, as Deleuze often reminded us, is not the same as making an onto-theological one between good and evil. The latter is what the techno-determinism of the western informational world does, fragmenting the event into neatly packaged but eminently consumable isomorphic spectacles of twittering teens and shady Salafists; those that were joyous around Anderson Cooper and those that punched him; ones that love America and ones that hate her. Techno-determined information flow is a form of power that seeks to reduce complexity into fungible data. It is there to destroy historical memory, to annihilate imaginative powers, to foreclose different emergent ways of thinking and being in the world. What it tries to abort at every step is a vision of alterity, a glimpse of a different world that is imminent.

Anustup Basu ‹basu1@illinois.edu› is Associate Professor of English, Criticism and Cinema Studies at the University of Illinois at Urbana-Champaign. He is the author of *Bollywood in the Age of New Media: The Geotelevisual Aesthetic* (Edinburgh University Press, 2010) and co-editor of *Figurations in Indian Film* (forthcoming from Palgrave-Macmillan in 2012) and *InterMedia in South Asia: The Fourth Screen* (forthcoming from Routledge, 2012). He is also the executive producer of *Herbert* (2005), which won the Indian National Award for Best Bengali Feature Film in 2005-06

FOOTNOTES

1 Thomas L. Friedman, 'Lessons From Tahrir Sq', *The New York Times*, 15 May 2011, http://www.nytimes.com/2011/05/25/opinion/25friedman.html

2 See Peter Hart, 'Friedman's Bogus Advice on Palestinian Non-Violence', 15 May 2011, http://www.fair.org/blog/2011/05/25/friedmans-bogus-advice-on-palestinian-nonviolence/. For more analyses and reportage on Palestinian non-violence, see for example Mary Elizabeth King, *A Quiet Revolution: The First Palestinian Intifada and Nonviolent Resistance*, New York: Nation Books, 2007; for rare journalistic reckonings see Mohammed Khatib and Jonathan Pollak, 'Palestinian Nonviolent Movement Carries on Despite Crackdown', 21 January, 2011, http://www.huffingtonpost.com/mohammed-khatib/post_1615_b_812459.html and Yousef Munayyer, 'Palestine's Hidden History of Nonviolence', 18 May 2011, http://www.foreignpolicy.com/articles/2011/05/18/palestines_hidden_history_of_nonviolence

3 See Patrick O'Connor, 'Nonviolent Resistance in Palestine', 17 October 2005, http://www.ifamericansknew.org/media/nonviolent.html

4 I have no intention of condoning these rocket attacks or other war crimes perpetrated by Hamas (including using the Palestinian population itself as a shield), and the idea of justice should not be understood in terms of symmetric violence. However, as documented by Human Rights Watch, such attacks caused only 15 Israeli civilian fatalities in the course of the decade. Hamas has often stated that the rockets were intended to hit Israeli military installations and not civilian targets. At times, it has expressed regret and 'sorrow' for Israeli civilian deaths. See 'Hamas "Regrets" Civilian Deaths, Israel Unmoved', Reuters, 5 February 2010, http://www.reuters.com/article/idUSTRE6143UB20100205. On other occasions it has pointed out a fundamental asymmetry in the issue itself when it came to the aggressor Israel, which has regularly bombed and shelled women,

children, and the elderly in Mosques, hospitals, and even schools. In the final analysis, the crudely manufactured Qassam and Grad rockets have no guidance system and have symbolically contributed more to the myth of Israeli insecurity and the concomitant spectre of Islamic 'terror' than achieved military objectives for Hamas. Sometimes the rockets have fallen short and hit Palestinians, as it happened on 26 December, 2008, for example, when a rocket hit a house in Beit Lahiya, killing two girls.

5   By December 2007, about 90 percent of factories and workshops in Gaza had closed down, primarily due to the lack of raw materials (Israel, at this point, was allowing only one-third to one-tenth of net requirements to pass through, including essential commodities like medicines, food, educational items, clothing, building and industrial supplies). Three-quarters of Gaza's population was surviving on $2 a day and a perhaps toxic water supply. About 70 percent of agricultural fields in the narrow strip of land that is 45km long and 8km wide had been laid to waste because there was an acute shortage of pipes and pumps required for irrigation, and also because Gazans were not allowed to farm in the 'buffer zone' designated by Israel along the border, which constitutes nearly one-third of the total arable land. Fishing has been forcibly restricted to three nautical miles from the coast, even though it should be 20 as per the Oslo Accords.

6   In comparatively recent history, perhaps the greatest sin of the Palestinians has been to elect Hamas to power, with a 56 percent mandate, in the Palestinian Legislative Council, on 26 January, 2006.

7   Consider Benjamin Netanyahu's recent rebuke to President Obama when the latter, following a long-standing American foreign policy position, suggested that a peace settlement be reached with Palestine based on the 1967 borders. That could not be done, said Netanyahu, because those borders have been rendered 'indefensible'. In other words, there might arise, in the future, the need for further strategic annexations in order to secure the now indefensible borders themselves. Following up on Netanyahu's assertion, Likud Party member and Deputy Speaker of the Israeli Knesset, Danny Danon, wrote an op-ed in *The New York Times* that suggested that, should Palestinians press for statehood through a United Nations General Assembly vote this September, Israel should pre-empt this process by completely annexing the West Bank, thus fulfilling a messianic tryst with destiny in the name of Greater Israel, and laying claim to the historic heartland of Judea and Samaria. This proposed annexation would of course be completed without extending citizenship rights to Arab-Muslims, who would remain a diminishing spectre of 'terror', now acutely cramped into areas progressively smaller than the 22 percent of their historic homeland, which is what the 1967 lines would have accorded to them. See Danny Danon 'Making the Land of Israel Whole', The New York Times, 18 May 2011, http://www. nytimes.com/2011/05/19/opinion/19Danon.html?_ r=2&partner=rssnyt&emc=rss

8   See Jacques Derrida and Bernard Stiegler, *Echographies of Television*, Cambridge, MA: Polity Press, 2002, p.4.

9   See Medea Benjamin and Charles Davis, 'Stop Arming Dictators', http://www. informationclearinghouse.info/article27753.htm

10   Linh Dinh makes this astute observation in 'Heartwarming Massacres from Iraq to Libya', 31 March 2011, http://www.commondreams.org/ view/2011/03/31

11   That is, without reminding the world of an essential calling of the Camp David accords, that Israel should withdraw its military presence from the West Bank. See http://www.jimmycarterlibrary.gov/ documents/campdavid/accords.phtml

12   See for instance Naomi Klein, 'Democracy Born in Chains: South Africa's Constricted Freedom', http:// www.naomiklein.org/articles/2011/02/democracy- born-chains

13   For an extended, insightful discussion, see chapter 1 of David Harvey, *A Brief History of Neoliberalism*, New York: Oxford University Press, 2007.

14   Ronald Judy, 'Reflections on Straussism, Anti- Modernity, and Transition in the Age of American Force', in *boundary* 2 33.1, Spring 2006, p.40.

15   One of the most significant cases of such 'information' extraction through torture was of course that of Ibn al-Sheikh al-Libi, who, under duress, made the false confession that provided the material for Colin Powell's notorious presentation to the UN Security Council to make the case for the Iraq war.

16   I of course allude to Marx's extraordinary tract *The Eighteenth Brumaire of Louis Bonaparte* in Karl Marx, *Surveys from Exile*, David Fernbach ed., New York: Vintage, 1974, pp.143-149.

17   The Egyptian military has been described as a sort of General Electric type conglomerate that

'virtually owns every industry in the country.' See for instance Alex Blumberg, 'Why Egypt's Military Cares About Home Appliances', http://www.npr.org/blogs/money/2011/02/10/133501837/why-egypts-military-cares-about-home-appliances?ft=1&f=2; and Tom Engelbert, 'Egyptian Math', http://www.commondreams.org/view/2011/02/14-2

18 Mubarak's last Finance Minister, Youssef Boutrous-Ghali, was an IMF alumnus who was chair of its policy advisory committee. He has been sentenced, in absentia, to 30 years in prison on corruption charges. Boutrous-Ghali was clever enough to leave Egypt before the heat got too much. See Wael Khalil, 'Egypt's IMF-backed Revolution? No thanks: Year after year, the IMF praised Mubarak's 'progress' signing up for its $ 3bn loan now hardly seems a break with the past', in The Guardian, 7 June, 2011, http://www.guardian.co.uk/commentisfree/2011/jun/07/egypt-imf-loan

19 I borrow this expression from the oeuvre of Paul Virilio.

20 See 'Tony Blair: Mubarak is not Saddam Hussain', http://www.politicshome.com/uk/article/21498/tony_blair_change_in_egypt_inevitable.html

21 See http://data.worldbank.org/country/egypt-arab-republic accessed 11 June, 2011 and page 13 of the ILO Report accessible at http://www.ilo.org/wcmsp5/groups/public/---dgreports/---dcomm/---publ/documents/publication/wcms_150440.pdf

22 See Amira Nowaira, 'Egypt's Day of Rage goes on: Is the World Watching?', The Guardian, 27 January 2011, http://www.guardian.co.uk/commentisfree/2011/jan/27/egypt-protests-regime-citizens.

23 The Pakistani newspaper The Dawn calculated in January 2010 that in 2009 alone, 44 predator drone attacks in the western tribal areas, especially North Waziristan province, killed 708 people, of which only 5 were certified terrorists. See http://archives.dawn.com/archives/144960. The going rate was thus 140 innocent civilians for every dead terrorist. The problem however is that some terrorists, like Illyas Kashmiri, reportedly killed in 2009 and then again in 2011, have a vexing habit of coming back from the dead.

24 See Walter Benjamin, 'The Storyteller: Reflections on the Works of Nikolai Leskov', in Illuminations. Ed. Hannah Arendt, London: Fontana, 1973, p.89.

25 See Eric Martin, 'World's Poor "One Shock" From Crisis as Food Prices Climb, Zoellick Says' at http://www.bloomberg.com/news/2011-04-16/zoellick-says-world-economy-one-shock-away-from-food-crisis-1-.html

26 See the World Food Situation Report at http://www.fao.org/worldfoodsituation/wfs-home/foodpricesindex/en/

27 Paul Buchheit points out that in 2008 the publication Price Perceptions said, 'index funds alone now own about 1 billion bushels of Chicago wheat compared to annual US production of about 500 million.' See Paul Buchheit, 'How Wall Street Greed Funded Egypt's Turmoil' in http://www.commondreams.org/view/2011/02/14-10

28 See the ILO report at http://www.ilo.org/wcmsp5/groups/public/---dgreports/---dcomm/---publ/documents/publication/wcms_149622.pdf

29 See https://www.cia.gov/library/publications/the-world-factbook/rankorder/2172rank.html

30 See Farah Marie Mokhtareizadeh, 'Over Wo(my)n's Dead Bodies: On Surviving "Liberation"' http://www.commondreams.org/view/2010/12/19 accessed 21 July, 2011. See the report itself in http://www.icosgroup.net/static/reports/afghanistan_dangers_drawdown.pdf. Another report by the The International Council on Security and Development shows overwhelming antipathy towards NATO operations: http://www.icosgroup.net/static/reports/bin-laden-local-dynamics.pdf

31 Arjun Appadurai, Fear of Small Numbers: An Essay on the Geography of Anger, Durham: Duke University Press, 2006, p. 36.

32 I am grateful to a group of brilliant colleagues and friends in India (Moinak Biswas, Prasanta Chakravarty, Rajarshi Dasgupta, and Bodhisattva Kar) who, in the course of a stimulating exchange of ideas across continents through Facebook, were of immense help in clarifying and conceptually enriching this trope for me.

33 See Alain Badiou, 'The Universal Reach of Popular Uprisings', http://kasamaproject.org/2011/03/01/alan-badiou-during-arab-revolts-the-universal-reach-of-popular-uprisings/

34 Karl Marx, Eighteenth Brumaire, op. cit., p.143.

THE
ILLEGITIMACY
OF DEMANDS

With demands over the wage and welfare in austerity Greece deemed illegitimate because unaffordable, what shape can struggle take? DEMETRA KOTOUZA *sees the all out attack on living standards as producing a de facto opposition that can't be cohered by ideologies of class*

With austerity escalating in Greece this year, there has been a parallel effort to resist it. Several strikes in key industries such as transport and electricity have taken place, mostly in the public sector, and six general strikes, accompanied by demonstrations of growing size and intensity. The 'indignants' direct democracy movement dominated attention in the summer, expressing parliamentary politics' legitimation crisis. In September, autoreduction practices became more frequent in response to new taxation, while universities and schools were occupied, the former against the new higher education bill and the latter triggered by delays in handing out books.[1] In October, a 48-hour general strike, with increased participation from the private sector, and accompanied by the occupation of most of the country's public services and infrastructure, brought everything to a standstill. Despite what was called by many 'the mother of all strikes' and the largest demonstrations in decades, which many thought might topple the government, the parliament passed a bill that essentially invalidates collective bargaining agreements and opens the way for wages to fall below the minimum. This sent the message that a large 48-hour strike is not enough to win a battle, and that worse is still to come.

This comes at a time when the struggle around the wage is becoming a matter of survival. Within a year, wages, even for those previously considered quite well off, have fallen below subsistence levels, to the point that paying bills, making rent payments and buying basics has become a widespread problem. This, combined with payment stoppages by employers, high unemployment and the decline of the petit bourgeoisie, as small businesses go bust one after another, is making survival the central question today, and the existence of the wage itself the most critical demand. However, it is not only this ruthless and abrupt attack on wages and labour rights, compounded by intensifying police repression, that makes these struggles particularly tough. Current struggles are facing a grim horizon, as the demands they voice are presented as impossible; even if small battles are won, it is unclear how winning the war would be possible when it is no longer fought at the level of a national economy, but rather in the midst of a global crisis with Greece as one of its epicentres. These battles are confronted with the risk of a default that could send shockwaves through the global financial system and bring about a wider recession and deeper impoverishment. To the extent that a default can indeed be triggered by the government's inability to implement austerity, these struggles appear to be self-destructive. But even if a default is inevitable, its prospect thwarts any hopes for a long-term victory that would make space for workers to go on the offensive. Facing this situation, it has become difficult to pose even defensive wage demands in a way that is effective and proportionate to capital's attack. The intense struggles that continue to take place have a feeling of despair, of hitting a wall.

This is not a condition that only characterises the class struggle in Greece, or even one that suddenly emerged in the current crisis. The global capitalist restructuring, which dismantles the social democratic institutions that guarantee survival for unemployed populations, began long ago. In so many ways it represents a return of the working class to its 'proper' condition, to its 'proper' entirely dependent relation to capital. Unemployment, both as a constant risk and a potentially long-term condition, as a constant underlying state of precarity integral to the condition of the working class, is becoming ever more prominent today. However, the current stage of crisis and restructuring is not a return to the situation existing prior to the birth of social democracy. The capitalist restructuring that began in the late '70s – characterised by the drive to reduce the cost of labour power through the development of advanced technology, the global zoning of production, and financialisation, with credit supplementing falling wages (up until 2007) to aid the reproduction of labour power in the western world – was a response to an earlier crisis of overaccumulation. The prospect of a renewed Keynesian 'deal', of a realignment of consumption with the wage, to 'productive' industrial capitalism, and the separation of national economies, is no longer possible because it is precisely what had to be done away with to overcome that crisis. Most importantly, the real subsumption of labour under capital has advanced to a level where there is no longer any possibility of a flight from capital for surplus populations as was the case with, for example, the creation of alternative, non-capitalist communes in the 19th century and Great Depression-era America. Class struggle is forced to address the capital relation itself, at the same time as capital denies the proletariat's role as the productive class which, as Theorie Communiste rightly argues, seriously undermines its ability to affirm itself within this antagonism.

This is confounded by the fact that there is no longer a singular, unifying working class experience that would generate a common identity on one side of the class struggle. The global and local zoning of production, and increasing precarisation, has fractured working class communities pushing, in the West, a large section into chronic unemployment and to survival through informal and illegal economies. In the global South, significant populations have been forced to emigrate to the West despite brutal repression.

In this moment of global crisis, this tendency manifests itself with great intensity in the 'second' zone of capitalist development and particularly in Greece. When even the demand for work cannot be satisfied at a broader, systemic level, let alone for the capacity of the wage to cover subsistence, even defensive wage demands appear structurally illegitimate whilst also being a matter of survival. The working class is having a hard time affirming itself as life – as labour power that needs to be reproduced – let alone as a productive force, in its relation to both capital and the state that used to guarantee its survival. The question of 'lost unity' also emerges as a central one, as conflicts within struggles intensify.

The contradiction between the necessity of the wage demand, and its lost legitimacy

---

Survival has become the central question today

reappeared in the indignants' direct democracy movement. The call for 'direct democracy now' rejected, in principle, the address of demands to a denounced political establishment and parliamentary system. It rejected dominant avenues of representation – the political parties and major unions – and put forward a call for

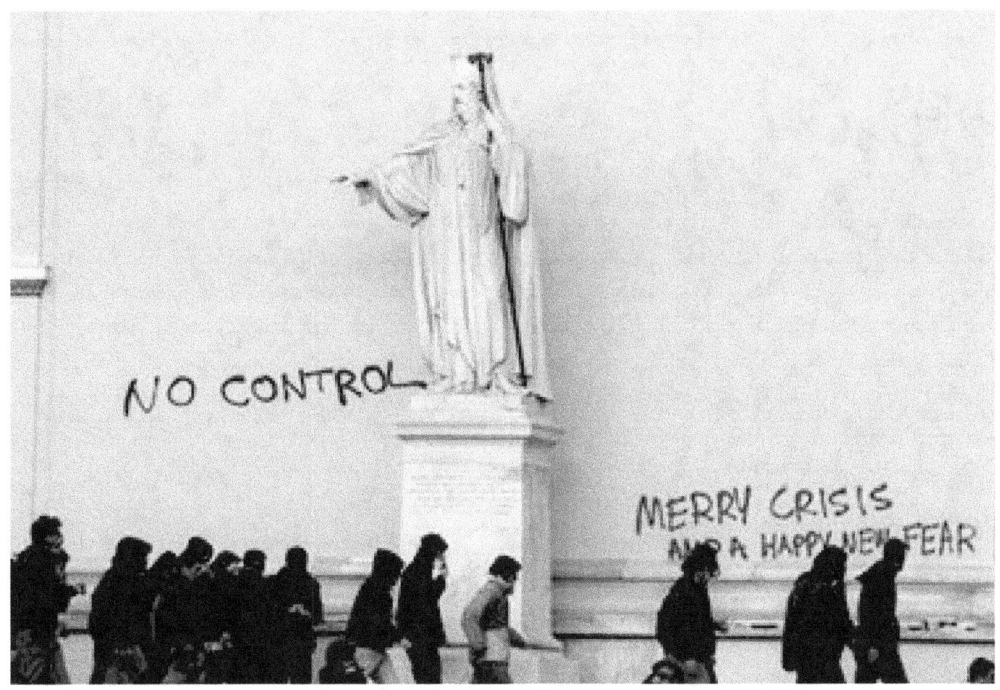

Anonymous, 'Merry Crisis and a Happy New Fear', Exarchia, Athens 2010

self-organisation: 'taking our lives into our own hands'. But, despite this language of autonomy, the movement was also driven by a single demand, namely that the Mid-Term Programme be voted down in parliament.[2] This suggests that building a defence within this face-off still takes precedence over any claim that it's time to self-organise and take over.

'Burn the parliament', the crowd shouted, but that did not amount to a rejection of politics. The direct democracy movement was clearly a political one, attempting to create a new politics from below, and even a political programme. Operating primarily at the level of political discourse, the 'direct democratic' imaginary envisioned a system of inclusive, bottom-up decision making, self-organised resistance and mutual support in neighbourhoods and workplaces. Similar to the indignants' campaigns in Spain and now the US, it was captivated by the notion that a more 'decent' life would be possible, if only the citizens had the political power. In the Greek case, the dominant conviction was that direct democracy alone, as a form of decision making, would be able to make capitalist production commensurate with meeting human needs, or, in its rather more militant version, that the democratic self-management of production would ensure those needs were met. The discourse of self-management, coming mostly from an alternativist anarchist tendency, and the broader conception of 'alternatives' – involving much speculation around alternative currencies and the autonomous circulation of agricultural products – sought to provide ideas for surviving the crisis or, less modestly, ways out of capitalism. However, all those ideas,

Anonymous, 'Rob Me', written on the shutters of a bank, Athens, 2011

beyond their historical limitations as practices, remain mostly just ideas, with the exception of creating a temporary self-organised enclave in public space. The attempt to develop immediate social relations within it (the rejection of money, a free collective kitchen, free lessons for homeless children) quickly reached the limit of an all encompassing capital relation (the return of money, closure of the kitchen because junkies and homeless 'took advantage').

Public political debate, which the direct democracy movement saw as its major strength, was also its limitation. The movement's dominant citizenist, democratic discourse was intrinsic to its inter-class character, explainable by the austerity measures' devastating effect on the petit bourgeois. Militants' attempts to push the discourse of class conflict came up against the principle of the 'people's' unity. In the midst of a relentless attack on the wage, debates around 'what is to be done' were muddled, unable to refer to a common class subject, whilst sporadic calls for a long-term general strike and other direct actions remained at the level of discourse. The assembly in Syntagma as well as those in neighbourhoods and towns around the country, mostly resorted to symbolic protests, public statements and expressions of solidarity, boosting or linking up existing struggles. They laboured to initiate actions other than the occupations themselves, which soon reached their own limits.

The imagined unity of national citizens against a failed system of government and decision-making also meant that immigrants were excluded by definition, except in the token action of inviting them to speak and organise events for a single day. Despite the active expulsion of extreme fascists from the Syntagma square occupation, the movement's citizenism was aligned with a growing nationalist anti-imperialist tendency, a response to the erosion of Greece's national sovereignty under the control of the Troika.[3] This provided the natural environment for a nationalist campaign against the Memorandum, the '300 Greeks', to set up shop, and for autonomous nationalists - who were in many respects unidentifiable so long as they kept quiet - to take part in the movement.[4]

Ridden with contradictions, the direct democracy movement experienced a fleeting moment of victory during the general strike of 15 June. That was a high point of struggle for the wider oppositional movement, with the PM almost resigning. The police repression and extensive anti-police clashes and rioting that took place during the strike, however, brought up renewed conflict within the Syntagma assembly, when the majority of its constituents rejected a motion that condemned 'violence in all its forms'. This moment was a major turning point that brought to a head the ongoing debate around proletarian violence. The direct democracy movement's relative tolerance of intense clashes with the police is not so much indicative of an anarchist influence, as of a wider tendency towards the use of such practices. Although these practices have been associated with anarchists, a growing

number of their participants are lower-class, precariously employed or unemployed youths – though the age range is broadening – who are more or less unrelated to the anarchist milieu. They accounted for a significant subsection within the direct democracy movement, to the extent that much of the assembly audience responded to conspiracy theories about 'violent agent provocateurs' by saying that 'the rioters are us'. After the defeat of 29 June, when the Mid-Term Austerity Programme was finally passed in parliament, rioting, as well as police repression of the demonstration, became

---

Citizenism has aligned with a growing nationalist anti-imperialist tendency, a response to the erosion of Greece's national sovereignty under the control of the Troika

---

exceptionally fierce, driving even more of those who had previously favoured 'peaceful protest' to change their minds. However this shift could not translate into practice at that stage. With the direct democracy movement weakened by its defeat, its internal contradictions combined with zero tolerance policing, a new round of struggles was anticipated.

The voting through of the Mid-Term Austerity Programme was followed by August's fast tracked vote on the new higher education bill that limits degrees to three years, flexibilises work contracts and rationalises university management, making further steps towards a business model for higher education. Importantly, it also abolishes 'university asylum' – the law that designates university grounds as off-limits to police – which has played an important practical role in social struggles since its institution after the fall of the junta in 1974. When students responded with occupations around the country after the start of the academic year, it already seemed too late. The peak of their engagement was in September, suggesting that the long occupations of 2006-7 may not be repeated this time.

Autumn also brought the rapid and ruthless slashing of indirect and direct wages in both the public and private sectors via cuts and emergency taxes. In response, auto-reduction practices spread more widely, having started a year ago in a more limited scale with the 'I Don't Pay' movement under the auspices of the leftist ANTARSYA party. The Public Electricity Company union refused to implement new taxation via electricity bills, bills were collectively burned outside tax offices, and there was a widespread tacit agreement that certain taxes would simply not be paid. The discussion around these actions again had, by its nature, an interclass, citizenist and legalistic character. Nevertheless, the fact that these were less symbolic political acts forming a response to governmental policy, but primarily acts of survival, as a large section of the population is unable to pay these taxes, links these campaigns directly to the crisis of the wage relation. With little room left for workers' struggles to develop around wage demands, these practices have temporarily claimed back a tiny fraction of the indirect wage, displacing the conflict outside the workplace.

The sense of despair in relation to winning demands, however, does not signal the end of wage struggles. When the government announced the impending abolition of the minimum wage and of collective contracts, as well as mass layoffs in the public sector, two general strikes were announced by the major

unions in October. That provided a basis for rank and file organising in workplaces to push for participation in the strike and for occupations in the public sector, especially in cases where they were met with resistance by management or by sectoral unions. Interestingly, although the entire public sector ceased to function for over a week due to mass occupations, these actions received public support, expressed in episodes such as residents blocking the way of strike-breaking private refuse collection vehicles – so much so that their drivers eventually went on strike themselves. The massive scale of the general strike of 19-20 October, and the emergence of rank and file organising at this juncture, does suggest that the struggle around the wage is what is driving social mobilisation in Greece right now. The staying power of rank and

---

Riots occur at the level of reproduction because this is exactly where the tendency of the wage to disappear is experienced

---

file organising, in spite of the general strike's inability to achieve its aims, is something to pay attention to. If their struggle escalates to the point that it challenges the official unions, but strikes and occupations are still not enough to win the fight to keep wages at a livable level, what type of practices might workers resort to?

The impasse of demands, the lack of prospects for even basic subsistence in a future of poverty level wages and high unemployment, combined with extreme police repression, does seem to coincide with increasingly forceful clashes at demonstrations, both against the police and between demonstrators. The

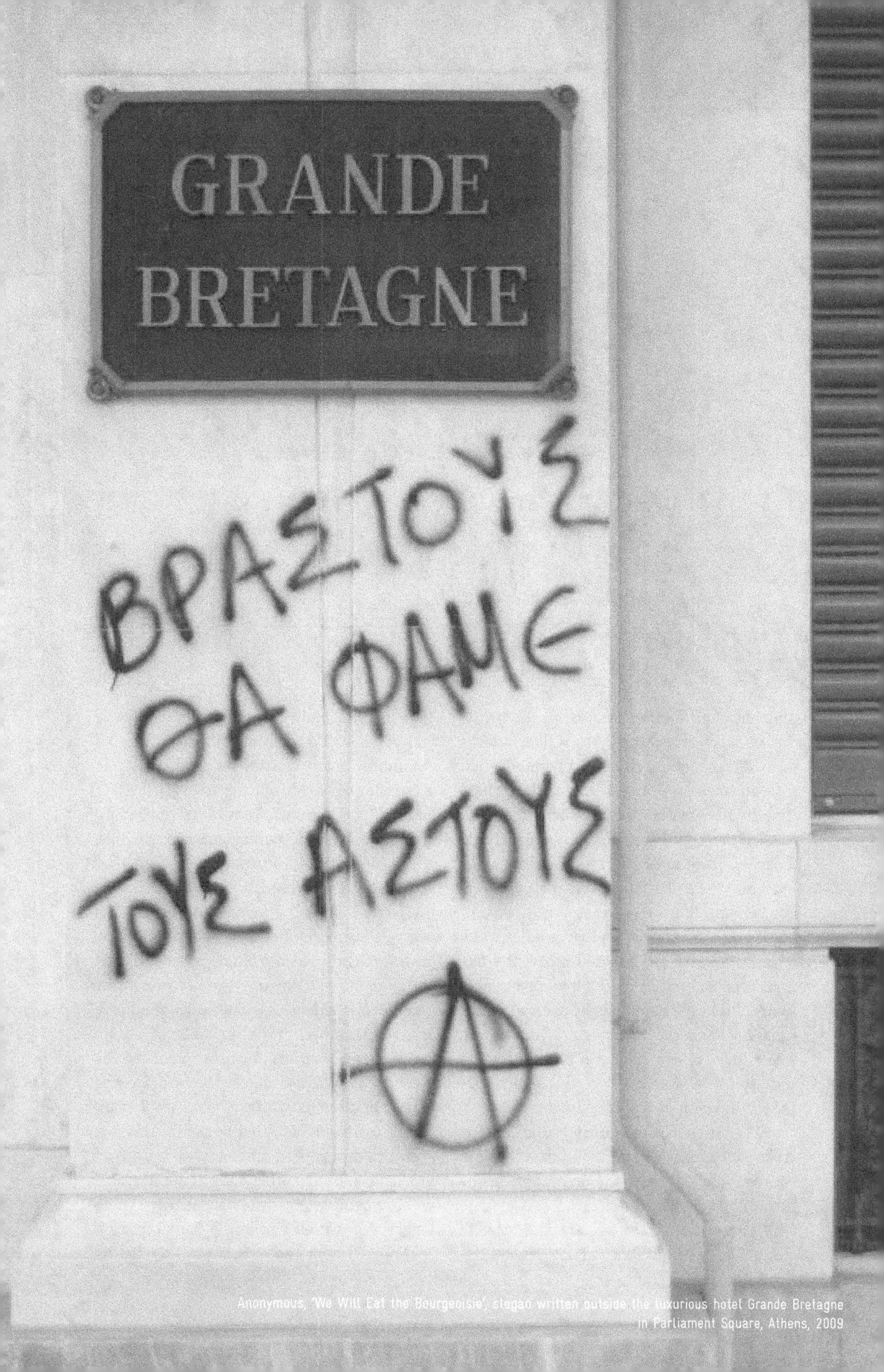

GRANDE
BRETAGNE

ΒΡΑΣΤΟΥΣ
ΘΑ ΦΑΜΕ
ΤΟΥΣ ΑΣΤΟΥΣ

Anonymous, 'We Will Eat the Bourgeoisie', slogan written outside the luxurious hotel Grande Bretagne in Parliament Square, Athens, 2009

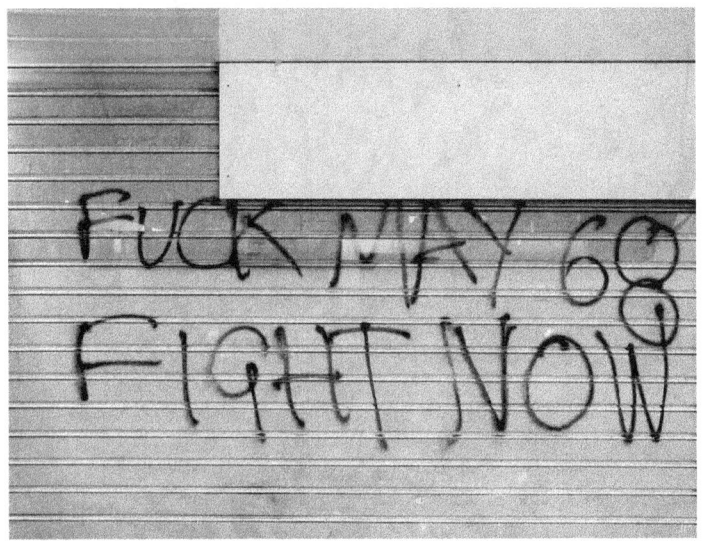

Anonymous, 'Fuck May 68, Fight Now', Exarchia, Athens

multiplication of direct attacks on the police, private and public property, as well as attacks on exchange and the obstacles to reproduction through looting – the latter fairly limited compared to recent riots in the UK and to those in Greece of December 2008 – signal that for many there are now zero stakes in social relations. Sustained attacks on the police are not 'missing the target'. They are in essence attacks on the enforced reproduction of social relations as they are imposed today. The fact that riots take place during national strikes suggests that they are a direct reaction to the contradiction faced by struggles around the wage. They occur at the level of reproduction because this is exactly where the tendency of the wage to disappear is experienced.

The serious clashes during the general strike of 20 October in front of the parliament, between the Communist Party union cadre (PAME) and demonstrators who had clashed with police the day before, are indicative of this tendency. On the second day of the most dynamic national strike and the most intense and populous demonstration in decades, the Communist Party played its traditional role of striving to lead workers' struggles while keeping them under control by encircling and protecting the parliament and its MPs, effectively replacing the role of the police. Other demonstrators attacked them as if they were the police, sparking a fierce street battle. This was not just a conflict about political tactics, however. As the Agents of Chaos have pointed out in a recent text, this was not a conflict between anarchism and Stalinist communism, as is often claimed.[5] It is a fundamental conflict between proletarian practices produced by the current cycle of struggles: on one side, the persistent attempt to affirm productive labour, to win demands within the capital relation, even the dream of 'taking over the means of production'; on the other, destructive practices without demands by living labour that can no longer affirm itself within the capital relation – a relation that no longer provides for its reproduction as labour power.

The current struggles in Greece contain within them the central contradiction continually produced in our time: the working class experiencing the limits of its struggle, which are its own intrinsic condition as living labour and the relations that constitute it as such. These struggles continue, despite the risk of a self-destructive outcome, namely a (disorderly) default. The threat of the destruction of capital, and with that the unavailability of work, does not stop struggles. This suggests that they could escalate in ways that break with the 'reasonable options' presented to them. Meanwhile, attacks on structures of social control, property and exchange, riots without demands and the inevitable conflicts they generate inside the struggles themselves, seem likely to intensify. It is the multiplication of these sorts of conflict, and not the triumph of productive labour and working class unity, that will shape the struggles to come.

Demetra Kotouza <demetra@inventati.org> is a contributing editor to *Mute*

NOTE

This text was written at a geographical distance from events. Many thanks to friends involved in the journal *Blaumachen* for providing invaluable information, ideas and feedback

FOOTNOTES

1   'Autoreduction' is the act by which consumers, in the area of consumption, and workers, in the area of production, take it upon themselves to reduce the price of public services, housing, electricity, taxation; or in the factory, the rate of productivity.

2   The Mid-Term Programme outlined cuts to services, wages, pensions and (what little remained of) benefits, and public sector layoffs, along with a long list of privatisations - the first step towards the total sell-off demanded by the 'Troika'. An interesting 'innovation' was that workers and pensioners were to be charged an extra 'solidarity tax' to pay for the one-year benefit given to the increasing numbers of the unemployed. Furthermore, it forecast that even after all these measures had been taken, by 2015 Greece's external debt would only have been reduced by a tiny fraction.

3   'The Troika' is a slang term for the three institutions which have the most power over Greece's financial future – or at least that future as it is defined within the European Union: the European Commission (EC), the International Monetary Fund (IMF), and the European Central Bank (ECB).

4   The emergence of autonomous nationalism and of frequent violent attacks on migrants by mostly working class far-right groups again occurs in the context of the fragmentation of the working class. Migrants are seen as the reason for the failure of wage demands, and in an attempt to regain bargaining power, a section of the working class that has lost hope in the demand for 'jobs for Greek workers', takes direct action to terrorise them out of the country, disregarding the laws of a sold-out government that is perceived as 'betraying' its citizens. However, the inability to unify as 'Greek workers' means that this tendency is very marginal despite its growth.

5   Agents of Chaos, 'Χωρίς εσένα γρανάζι δε γυρνά...' ['Without You, Not a Single Cogwheel Spins'], October 2011, http://athens.indymedia.org/local/webcast/uploads/xwris_eseva_gravazi_de_gurva.pdf

WILL CHINA SAVE GLOBAL CAPITALISM?

All over the world, the capitalist states are taking austerity measures to slow the growth of their debts. It is obvious that this policy, since it slows consumption, can't in itself sustain the growth required for capital accumulation. Where, then, can the necessary economic stimulus come from? For lack of alternatives, eyes are turning eastward. It seems, writes SANDER, *that history, the supreme ironist, has chosen 'communist' China as global capitalism's saviour.*

All over the world, the capitalist states are taking austerity measures to slow the growth of their debts. It is obvious that this policy, since it slows consumption, can't in itself sustain the growth required for capital accumulation. Where, then, can the necessary economic stimulus come from? For lack of alternatives, eyes are turning eastward. It seems that history, the supreme ironist, has chosen 'communist' China as global capitalism's saviour.

WHAT CRISIS?

The current crisis is the result of the obsolescence of the very basis of the capitalist mode of production, the value-form. It is the value-form which forces capitalists to continue to use abstract labour time to measure wealth, while the creation of real wealth has become less dependent on the amount of labour time used than on general knowledge and its application in production. Marx's prediction (in the *Grundrisse*) is fully realised today. It is in this developing contradiction that he saw the historical limit of capital. It has become absurd for humanity to base decisions on what to produce, how, how much, where and for whom, on the law of value. This absurdity manifests itself in the simultaneity of generalised overproduction and extreme poverty, in the increasing incapacity of capital to exploit the labour power at its disposition, causing an accelerating expulsion of workers from production, while money seeks

a false security in financial bubbles. But capital is subjected to the law of value like an animal is subjected to its nature. It cannot solve a problem whose solution calls for its abolition. It therefore can do nothing against its crisis except fight its symptoms, blow hot and cold, alternate stimulus measures and austerity measures and delay the inevitable descent.

THE METAMORPHOSIS OF VALUE

The accumulation of capital is going through cycles in which value morphs from money into commodities and then back into money: M – C – M'. Money, M (abstract value), is the starting point. It buys commodities, C, the means of production whose value is transmitted in the commodities resulting from their productive use. These new commodities are sold, which transforms the value back into money, M'. The only reason why the initial money, M, was transformed into C, is that M' (or M prime) is greater than M. The transformation is profitable.

Marxist analysis reveals that the source of profit is surplus value, the difference between the value of the living labour power that the capitalist buys (which, as for all commodities, is equal to the quantity of abstract labour necessary to reproduce it) and the value it creates for him (the quantity of abstract labour performed). The higher the productivity, the less labour time is needed to produce the equivalent of wages, thus the greater the part of the workday that produces surplus value. But this surplus value can never arise from more than a part of the workday. The technological development which increases productivity also decreases the value of living labour in production relative to that of past labour (technology, infrastructure). Of this living labour, surplus value is only a part and it therefore must decline with it. Since profit = surplus value, this is a problem, especially in a world that operates more and more on automated processes. Productivity does not save capitalism, on the contrary, it ripens and

---

The neoliberal arrangement counteracted the tendency to overproduce by giving money other destinations than productive investment

---

further accentuates its contradictions. The more it increases and the more these increases become widespread, the more the value of what is produced declines relative to the value of the capital invested in production.

It creates another problem in the next phase of the cycle of value, the transformation of commodities back into money, C – M'. This does not happen automatically. The increase of productivity slows the production of value, but accelerates the production of use values. Unproductive consumption can always be expanded, but productive consumption remains limited to the use values needed for production. These do not increase because the ability to produce them increases. The essential market consists of the demand for capital goods and consumer goods necessary for the reproduction of labour power. It's their expansion that makes the expansion of value in the next cycle possible.

It's this market that, over time, is incapable of following the acceleration in productivity. The general overproduction of technology (visit cities like Detroit, if you need proof) and of labour power (nearly 2 billion unemployed) testify to it.

NO VALUE WITHOUT A HOARD

When these bottlenecks reappeared in the 1970s, after 'the 30 glorious years' made possible by the war and the expansion of the global market under the aegis of the dollar, the general tendency was to inflate, to support demand, to stimulate M – C. The law of value punished this cheating with accelerating inflation.

Attempts to get it under control by forcing the working class to shoulder the burden faced intense resistance. The growth of fictitious capital in the circulation of commodities devalued money and made it seek refuge. It discouraged M – C, productive investment, and encouraged speculation. M preferred to stay M, instead of transforming itself into commodities. But it couldn't.

Capitalism cannot survive without a 'treasure'; money must be able to leave circulation without losing its value to be re-injected at the right time. But money, abstract value, is not stable. Its power lies in its ability to transform itself into other commodities. Therefore the value of the monetary hoard remains dependent on real valorisation, on value creation, which can only happen in phase C, in production. Otherwise, it becomes paper or less. Inflation signals that this valorisation decreases relative to the money in circulation. If hoarded, money is dragged down by the loss of value of money in circulation and panic ensues. Accumulation loses its purpose. Money desperately seeks refuge in gold or old paintings and tries to protect itself with exorbitant interest rates that are strangling an already crippled production... it's one of the possible paths to breakdown.

Value is an objective abstraction, that is, a social construction that has taken on the appearance of being objective, to be an intrinsic feature of things. It is not. In the end, it is a belief system that collapses when money cannot be hoarded.

The restructuring of capital that began in the 1980s brought inflation under control, boosted the rate of surplus value and thus the rate of profit, and restored confidence in hoarding. In other texts we have analysed in greater detail how this was done.[1] Amongst other things, we pointed to the crucial role played by globalisation: the global integration of production chains and markets, deregulation and globalisation of financial capital, the emergence of post-Fordist production in advanced countries and the massive displacement of Fordist industry to low wage countries.

CHINA TO THE RESCUE

China was by far the country most changed by this restructuring. In a few decades, it transformed from a failed attempt at autarkic state capitalism into the second largest economy in the world and the largest industrial producer. In 1990 it produced 3 percent of the world's industrial output; 20 years later 19.8 percent, overtaking the US who had held that position for 110 years.[2] China's dramatic expansion has benefited the advanced capitals in several ways: its cheap products were the main reason why inflation remained low, the combination of its low wages and modern technology brought huge profits to Western and Japanese investors, and the realistic threat of moving production

to China helped to curb wages in the advanced countries. For the expansion of the world market, its impact was also crucial: less by the opening of its domestic market (which is large and growing, but limited by the extreme poverty of the majority of its population) than by its indirect effect on the market of its customers. Because its expansion was driven by external trade, and because the state kept the lid on Chinese wages and thus on the consumption of the working class, since their low level is its main competitive weapon, each year China obtained a growing trade surplus. As in other countries before it (especially Japan), whose industrial development depended on the US market, China used these profits to accumulate a hoard consisting of dollars, public debt and US securities.

By hoarding these dollars, China withdraws them from circulation, and thereby keeps the dollar stronger than it otherwise would be. That's the main reason why China does this: to defend its competitive position in the market towards which its industry is essentially oriented. For the same reason it buys American public debt, thus giving the Fed the means to stimulate demand by lowering interest rates. China's strategy, whether it likes it or not, is based on its confidence in the US dollar as the guardian of value.

By selling commodities under the value they would have if they were produced locally, and by accepting a payment that is largely hoarded instead of demanding an immediate equivalent, China, and other countries in a similar position, not only directly stimulate the purchasing power of their export markets, but also do so indirectly by facilitating an inflation of their assets. American capital led the dance. With interest rates approaching zero (which wouldn't have been possible without the demand of China and Japan for its debt,) tax giveaways, deregulation, privatisation, the commodification of services and finances, it inflated the demand for its real estate and securities and thus their price. The trust in the capacity to hoard value was fully restored. In 2004 the economist Stephen Roach estimated that 80 percent of the net savings of the world flowed to the US. A growing part of the global profits were siphoned away from general circulation into the American hoard. After the crisis erupted, the 'neoliberal' policies which had stimulated this arrangement came under heavy fire, since the crisis had revealed its speculative essence. But what was the alternative from a capitalist point of view? The measures that should have been taken according to the capitalist left – more productive investment, if necessary directly by the state, and higher wages to stimulate demand – surely would have meant that the threats of overproduction and accelerating inflation would have returned much sooner.

The neoliberal arrangement at least had the advantage of holding back these threats for a while. It counteracted the tendential overproduction by giving money other destinations than productive investment. It counteracted inflation by sucking money out of general circulation. And it made the rich even richer – especially the traders in money and everything that can be easily monetised. 'The real profits are not made by producing,' said a Wall Street man, 'they're made by buying and selling.' Or even by doing nothing, since the prices of shares and real estate rose every day. It became quite rational to go into debt, since the rise of 'values' more than compensated for the low interest obligations – if you had money. If you didn't, it was still expensive to run up debt; but for the rich, it paid for itself and then some.

No surprise then that the illusion took hold that capital can accumulate in the form M – M', without having to pass through that annoying phase C.

But in reality, it is only in this phase that value is created; that the value invested in means of production, C, and labour power, V, transforms into C+V+S (surplus value). Thanks to the inclusion of China and other low wage countries and thanks to the relative decline of wages in the advanced countries, the creation of value grew, but not at the dizzying speed of the hoard.

The value of the hoard is not an objective fact but an article of faith. To defend the faith in its hoard is the primary objective of the capitalist state. That is the faith for which the crusades of our days are waged: to project power; to reassure the shareholders.

THE FALSE PROMISE OF AUSTERITY

When the crisis pierced the bubble and showed that the apparent enrichment was to a large extent due to the insertion of fictitious value in the cycle of value, the capacity to hold value once again became doubtful. It took an historically unprecedented acceleration of spending, and thus of debt creation, on the part of the strongest countries to support the financial institutions, to avoid a collapse of faith in the private hoard. Faith in the state is what saved them. But, to confront the consequences of the growth of fictitious capital, much more fictitious capital was created. And it continued. With its 'quantitative easing' policy the American Federal Bank continued to support the prices of public debt and mortgages by buying them from the banks with money it created out of thin air. The European Union created hundreds of billions of euros to save its most indebted member states from bankruptcy. Even the countries where draconian austerity measures are imposed didn't stop creating debt. They can't function without it; at the very least they need to re-finance their old debts. None of them has a budget without a large deficit. So public debt keeps swelling, while austerity undermines demand and therefore also the receipts of the state so that more debt must be created... in this way, the crisis of confidence in the capacity of private capital to hoard value is transformed into a crisis of confidence in the state as guardian of value. This crisis already severely affects the weakest competitors and is moving towards the centre of the system.

Those trillions of new debts are commodities which must compete with all other commodities to find buyers. Their growing supply demands a growing portion of the purchasing power so less remains for other commodities; this increases the saturation of markets, which discourages productive investment and thus the creation of new value.

Austerity serves to improve the brand image of the country, to inspire trust in its future ability to pay its debts. The growth of public debt means that the competition between them for capital is intensifying on the basis of that trust. The larger the supply of debt of so-called 'save havens' like the US, the more countries whose debts are less trusted are forced to improve their ability to pay with austerity measures to avoid becoming the victim of capital flight.

So the goal of austerity is to convince the capital markets that it is profitable to buy its public debt; that its capacity to hoard value remains intact. But this strategy remains based on the illusion that M can become M' without an expansion of value in the C phase. It bets that the economy can pay for exponentially growing debts without a corresponding growth

of production. It's a short term strategy: the savings create space to pay the creditors but they reduce the creation of new value and thereby reduce the future capacity to repay debts.

In the sphere of production, the emphasis is on cost reduction as well: savings are made on employment, wages, materials and

---

China's 150-200 million internal migrants came from its vast interior, in a well orchestrated exodus, to provide the labour power for the 'global assembly line'

unproductive costs. These savings, especially the first two, have made the recovery possible. In this recovery, however, the lost jobs have not come back: more is now being produced by fewer workers than before. This reflects an increase in the rate of exploitation (S/V), but also an increase in the organic composition of capital (C/V). This was not a result of a boom in technological investment. A reduction of V (labour power) was already technically feasible earlier but it took the excuse of the crisis to impose it. This trend further diminishes the demand for consumer goods on the part of the working class, thereby sharpening the problem of the realisation of value; and it diminishes living labour in relation to past labour in production, sharpening the problem of the creation of value.

For capital, there is just one way to defend itself against the devalorisation that the law of value demands: make the working class pay for the crisis. But the unprecedented wave of strikes in China and other Asian countries last year, the Arab Spring, the resistance against austerity by the proletariat in Greece and other countries,

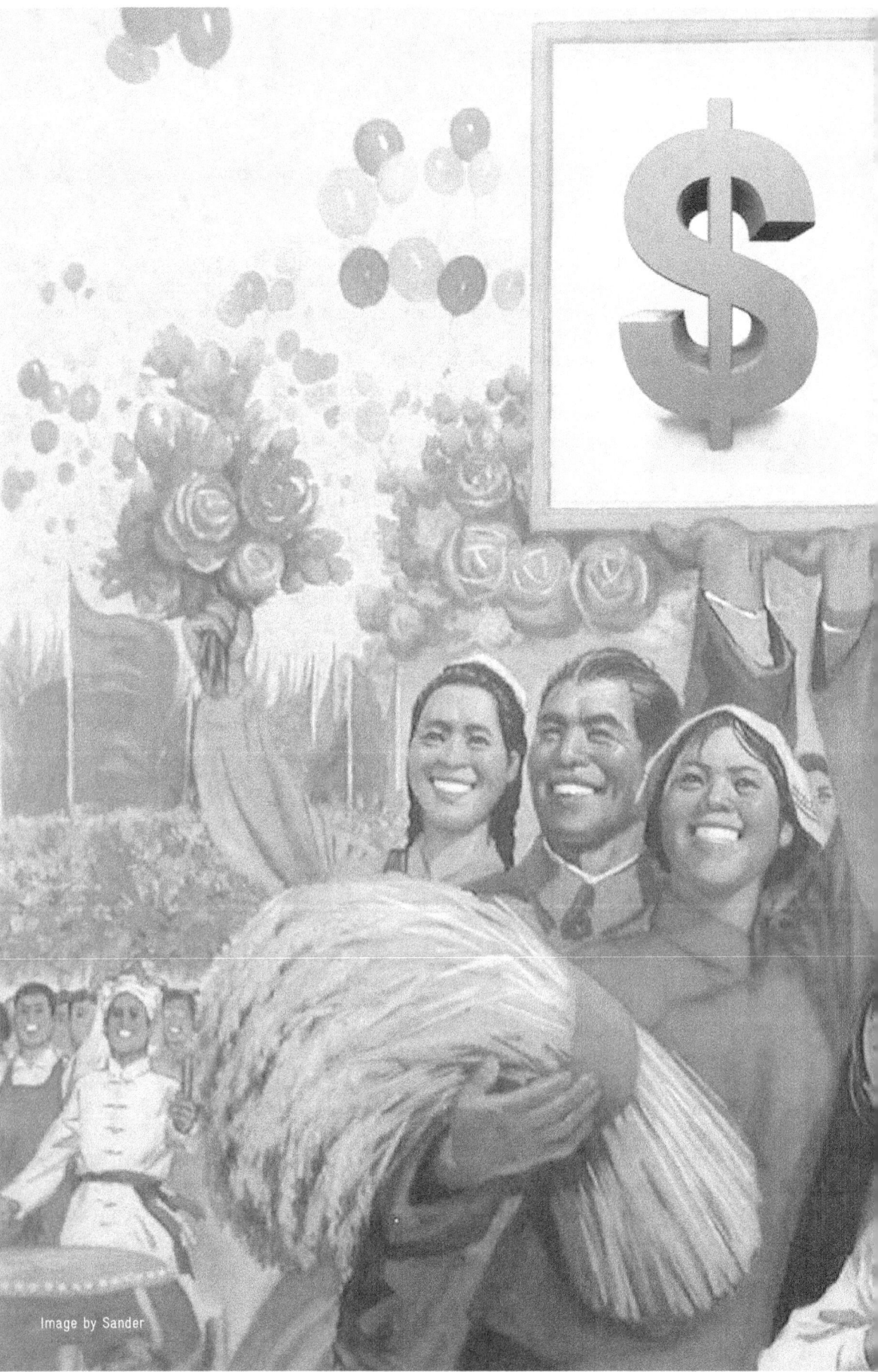
Image by Sander

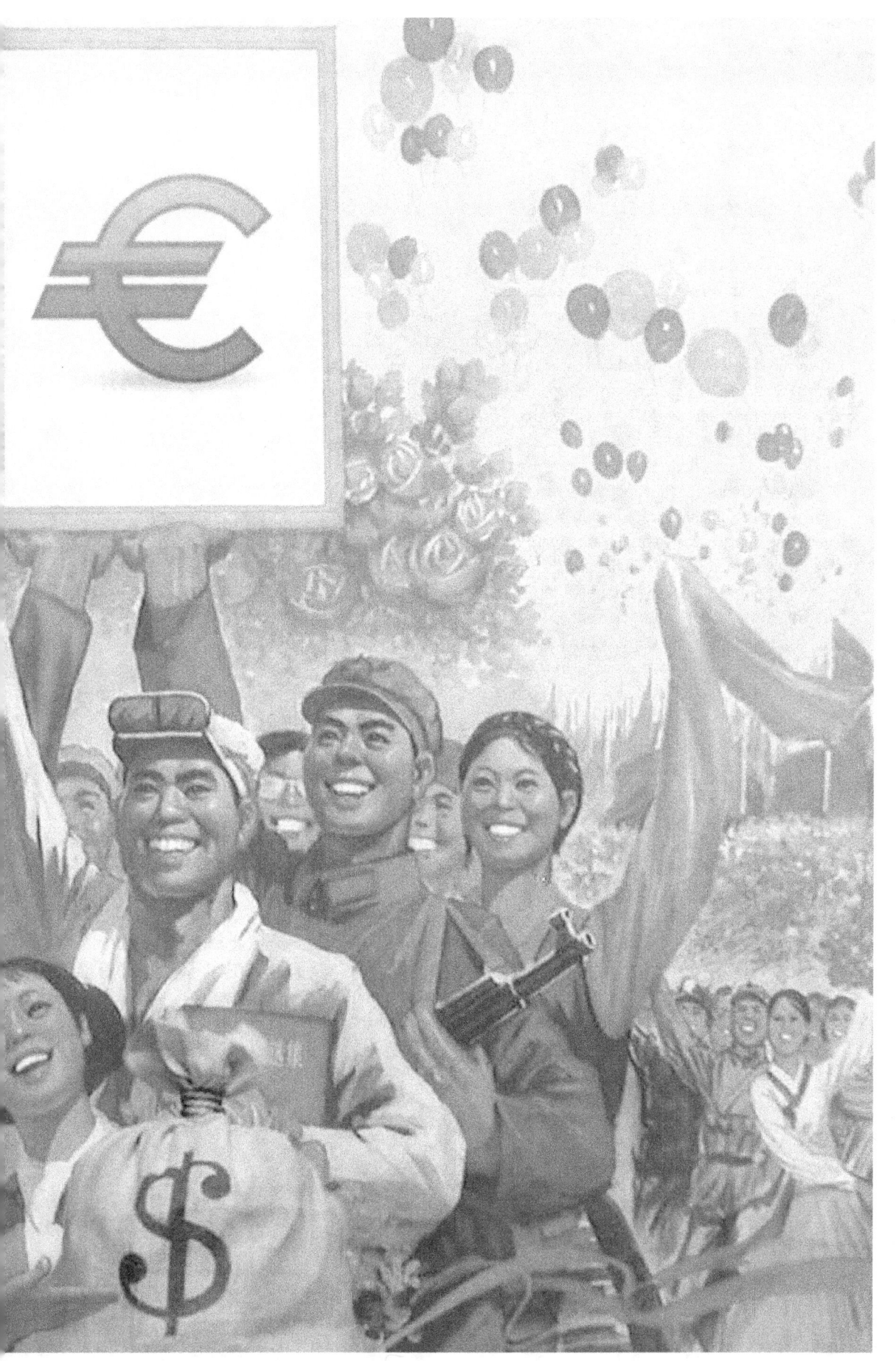

show that this will become increasingly difficult – and risky too. States are constrained by their fear that a point will be reached where social control escapes them. Already, young proletarians who occupy public spaces in Spain, the US and elsewhere, are beginning to wonder whether another world is possible than the world of value.

But for capital, there is no alternative. None of the scenarios that its apologists invent offer an escape from the iron cage in which the law of value imprisons it. In other texts, we analysed why 'green technology' will not save it, and why information technology, monopolisation and artificial shortages will not save it.[3] Then there is the hope placed on China. China seems rich and in dire need of just about everything: the perfect market to revitalise the global economy.

Will China save capital from drowning? To a large extent, it already has done so during the last quarter of a century, as we saw earlier. But evidently, its beneficial effect for global capital has not prevented capital from descending into its worst crisis since the 1930s. So to get it out of this crisis, this beneficial effect would have to increase. But the opposite is happening. Both as a source of surplus value, and as a market, China's beneficial effect is diminishing; the former because of the rising value of labour power, the latter because of its own growing indebtedness and inflation.

THE RISE IN THE VALUE OF LABOUR POWER

China's beneficial effect was primarily based on its abundant supply of dirt cheap labour power, well disciplined with the help of Confucius and Stalin. It is weakening because the development of China has changed its society and this is pushing the value of labour power higher.

The majority of the workers who make all these cheap products that keep inflation down in the West are migrants (that is the case for 80 percent of the miners, 70 percent of the construction workers, 68 percent of the industrial workers and 60 percent of service employees). They are between 150 and 200 million strong and they came from the vast interior of the country, in a huge but well

---

China becomes an exporter of green technology, while its factories vomit poison into the air as if there were no tomorrow

orchestrated exodus, aimed at providing the labour power for the 'global assembly line.' The first generation of migrants consisted of peasants and other villagers who never knew anything else but a world of poverty. The value of labour power is determined by its cost of reproduction, but this differs from one society to another. In the interior of China, as in India, where the society has been characterised by general poverty for many generations, the consumer goods that are considered socially necessary for the reproduction of labour power are minimal. That's what makes the value of its labour power so low for capital.

The way in which Chinese capital has managed the labour force shows that its aim was to prevent this from changing. For this, it used the *Hukou* registration system that ties the worker to the place he or she comes from. That means that the migrant worker has no right to benefits such as health insurance, except 'at home' (where they often don't exist), no right even to stay when he or she becomes unemployed. There is a strong resemblance to the 'homeland' system under South Africa's Apartheid regime, and with the treatment of undocumented workers everywhere. The *Hukou* system is designed to meet several objectives: the artificial determination of the value of labour power on the industrialised coast by the conditions of the hinterland; the creation of division within the working class; making workers vulnerable to intimidation; and preventing the migration from the interior becoming an avalanche.

The sons and daughters of the first generation are still considered 'migrants' under the *Hukou* system, but they live in a different world than their parents and have fewer links with their place of origin. They are urbanised young people who live in an environment that is much more technologically developed, complex and rich. An environment that is also transformed by the extravagant consumption of all those newly rich they see around them.[4] The emergence of an industrialised society implies a change in the value of its labour power: the consumer goods seen as necessary for its reproduction inevitably expand. The young generation no longer accepts the *Hukou* system and the conditions that stem from it. Because of this pressure, the system was already decomposing and the strike wave of last summer dealt another blow. Wages were already rising considerably in the industrialised coastal regions, even for migrants (between 2003 and 2009 by almost 80 percent). And it has continued. In the last two years wages in the coastal regions rose by 50 percent.

There are already capitals that are leaving these regions to set up shop where wages are still lower, as in Vietnam or Bangladesh, or in China's interior. But there too, the changes in living conditions resulting from industrialisation are pushing wages higher.

Furthermore, the growing combativity of the Chinese proletariat has had an impact on the consciousness of workers in the region. In Vietnam and Bangladesh, the number and intensity of workers' struggles has shot upwards since 2010. Borders are less and less capable of preventing such contagion. News travels fast outside of the controlled media. In Vietnam wages are rising as fast as in China. In Bangladesh the minimum wage was increased by 85 percent last year. In China's interior wages are still considerably lower than in the coastal provinces, but they are rising at a faster pace.[5]

So it appears that capital's capacity to combine modern technology with ever lower wages, which sustained its rate of profit for at least two decades, has reached its limit. It is true that there are still places on earth where the value of labour power is lower (in particular in India) but there, other factors, such as the lack of infrastructure (roads, ports, power, etc.) poses severe limits. So the hope that the cheap labour power of China and similar countries will revitalise global capital is not based on perceivable trends in the real economy. It is true that this could change if Chinese capital were to succeed in pushing the price of its labour power far under its value, but for the moment conditions are not in its favour.

THE STIMULUS POLICY CREATED A BUBBLE

The vertiginous growth of its exports in the past decade made it possible for China to reduce the share of wages in the GNP dramatically, whilst conceding a rise of wages at the same time. The expansion of the pie was large enough to accommodate a growth of the purchasing power of workers, even though wages became a smaller part of the pie. Today, that is no longer the case.

The Chinese economy, as it is structured around its export sector, obviously suffered huge losses when its markets shrank after the crisis burst open. The state, concerned about the social consequences of a slowdown of the economy, reacted with an ambitious stimulus programme. Only the US spent more. But while the US created money to back up its treasury, American assets, China did so in the first place to stimulate investment. State-run banks unleashed a lending spree which led to a construction boom of unprecedented proportions. Spending on fixed asset investment is now equal to nearly 70 percent of the nation's GDP. It is a ratio that no other large nation has approached in modern times (in the US, the figure has hovered around 20 percent for decades). But did this growth of credit lead to a corresponding growth of value? Apparently not; more and more debts are no longer paid off. The money that was created in their name is fictitious, yet it circulates. Debt, speculation and inflation are forcing China to end, or at least sharply reduce, its stimulus policy. The hopes of those who see in China a market that will continuously expand will be rudely disappointed.

China's stimulus measures have helped significantly to soften the crisis of advanced capitalism. When China buys, day after day, billions of dollars, it gives the Fed the flexibility to create more money. China does it to curb the devaluation of the dollar vis-à-vis its own money, the Renminbi (RMB), also known as Yuan, in order to protect its competitive position on the American market. More precisely, many companies in the coastal provinces which produce for the external market already operate with a razor thin profit rate. Their contracts are in dollars, but they pay their suppliers in RMB. A sharp devaluation of the dollar would be a fatal blow to them.

So it is not surprising that China used its stimulus programme to reduce its dependence on Fordist production, by trying to become a producer at the cutting edge, where profits are less derived from the low value of labour power than from technological rent (i.e. a market advantage). The efforts it has made towards this goal, such as the modernisation of its infrastructure, were beneficial for the exports of the advanced countries, especially for Germany, the leading producer of modern technology. From a country with antique trains, China became an importer of HSTs (high speed trains). But now it is becoming an exporter of HSTs. That changes the game. The privileged sectors are becoming crowded. China becomes an exporter of green technology while its factories vomit poison into the air as if there were no tomorrow.

A CURSED TREASURE

At first sight, it seems so simple. China has huge needs and huge financial reserves. Just do the maths and everyone benefits. But it is only simple if you think money and value are the same. If China decides to become a cutting edge producer in all areas, using its financial reserves to buy the best technology in all sectors, it would be for a time – before this ends in gigantic overproduction – a huge market of last resort for the rest of the world. But it can't.

These financial reserves are the debts and money of other countries, especially the US. To what degree does this money represent real value or only fictitious capital? Its partially fictitious character remains hidden in the hoard as long as the faith in it remains intact, but it would appear clearly as soon as China attempted to bring the amount of dollars needed to realise that plan into circulation. By taking huge reserves of dollars out of the sterility of its coffers where they can do no harm, to throw them into the global economy, China would achieve the opposite of what it wants: a sharp devaluation

---

It's as if there was a curse on China's hoard: these trillions of dollars will keep their value only as long as they remain untouched

of the dollar, which would destroy the profit rate of its export industry and would devalue its own financial reserves, with a worldwide acceleration of inflation to boot. And if it were to use its hoard not for massive investments but to finance a general rise of purchasing power, the consequences would be equally catastrophic: the price of labour power, which remains its main competitive weapon, would shoot up; inflation and speculative investment, which already have reached alarming levels because of the accelerated monetary creation, would become unstoppable. It's as if there was a curse on China's hoard: these trillions of dollars will keep their value only as long as they remain untouched.

THE FEAR OF A HUMAN AVALANCHE

There is another reason why China cannot make this 'easy sum'. It could, for example, raise its agricultural sector to a point where it would be as productive as that of the US. Technically, nothing stands in its way. But instead of doing this, China is prospecting Africa and Brazil to buy land to start modern capitalist farms there, far away from home. Because doing this at home would mean the expulsion of hundreds of

millions who would flee to the cities. That is the social nightmare that the ruling class wants to avoid at all costs. The same is true in many other sectors. China can't be reduced to an industrial zone in the south and subsistence farming in the interior. The majority of companies, employing a majority of the working class of the country, are capitals of a low organic composition; that is, employing lots of workers but at low productivity. They have survived, thanks to the low value of labour power (reinforced still by Maoist rule, when the value of labour power meant enough to survive just until tomorrow, the 'iron bowl' and nothing more), and thanks to the fact that China's internal market is only partially opened to outside competition. But also thanks to loans from the banks, that is to say, from the State.

During the last three decades, the State has stopped supporting thousands of those companies. This not only in order to cut expenses but also to feed – not too much, not too little – the stream of labour power needed by the rapidly expanding Fordist industry in the South. But there are still millions of them. To support them was the main goal of China's stimulus programme. It did so by giving them orders (infrastructure projects, of which a principal goal is to be able to move large numbers of migrants to and from the industrial zones) and giving them loans of which a large part will never be repaid.

According to the IMF, China's rate of debt to GDP is 22 percent, a lot lower than that of the US or the EU. But this figure does not include the debts of the thousands of investment companies formed by local governments that invested in infrastructure and in companies that, from the point of view of value, no longer have a reason to exist. According to the calculations of Victor Shih of Northwestern University, the debts of these investment companies amounted to 11.4 trillion RMB ($1.7 trillion) by the end of 2009, or 35 percent of China's GDP. Taking into account the open credit lines already assigned to them, they would rise by another 12.7 trillion RMB by the end of this year, to a total of $3.7 trillion. Already 28 percent of these loans are 'non-performing'. 'Most of the government entities that borrow can't even make the interest payments on the loans', Shih said to *The New York Times*. When these debts are included, China's rate of debt/GNP was 75 percent at the end of 2009 and would be 97 percent by the end of 2011 – higher than that of the US today (94 percent). So the hope resting on the assumption that China can play the role of locomotive because it doesn't have to carry excessive debts which force the other economies to austerity policies, seems unjustified. In China too, the State's capacity to compensate for the lack of value creation by creating debt, is eroding. Its efforts have led to excessive indebtedness, an inflation rate (officially six percent, in reality at least double that) that threatens its capacity to monetise value, and speculative bubbles, especially in real estate.[6]

For these reasons, China had to end its stimulus programme. Since last fall, the State has ordered the banks to drastically tighten their loans, and has begun to consolidate – i.e. liquidate – thousands of local investment companies. Its economy is beginning to cool. At the same time, the pressure for higher (or less low) wages continues. Workers in transportation, services and white collar jobs who did not get the raises that the industrial workers obtained, claim it's their turn now.

Some China watchers think that the measures China is taking are too little and too late to get inflation under control; that a climate of 'stagflation' will make it very difficult

for the State to maintain its grip on society. It is not up to us to predict whether China will make a 'hard' or 'soft' landing. But in both cases, the high hopes invested in its market will land hard.

China will continue to grow, but less than before. Like elsewhere, this growth will create fewer jobs than it will destroy. Out of fear of social convulsions, China tries to limit this tendency, but this is becoming increasingly difficult. In China, like elsewhere, the great worry of the ruling class is: how will we manage all this superfluous variable capital? Not just the migrants and other refugees from rural poverty, but also the millions of graduates for whom there is no more place in the economy.[7]

Everywhere, the nightmare of capitalism becomes: what will we do with all those people? Where can we stockpile them, how can we keep them quiet? How to separate the superfluous from those we need? How to prevent them from engaging in revolt? How to make them disappear?

For the moment, capital is focusing on reducing its costs. It is well aware of the impossibility of creating new debt to replace the old non-performing debts endlessly, or, in other words, of the impossibility of continuing to hide, with fictitious capital, that the capital in the hoard is (to a growing extent) fictitious. So, by reducing its costs, it seeks to create the financial space to defend the confidence in the capacity of its debts to hold value. In the past three years, trillions of dollars, euros, yens and RMBs have been created to support the private hoard undermined by bad debts, and trillions more to impede a deflationary spiral towards depression. Never has so much money been created in so little time. This has put a brake on the deflationary pressure without eliminating it. It remains a bubble economy. Capital, M, continues to skip the phase C to get to M' and, by doing so, undermines M'. All the money creation, the tax breaks and other presents to the possessors of capital can only hide this for a limited time.

So the pendulum swings from stimulation to austerity. China is ending its stimulus programme, the US has ended its 'quantitative easing' policy and, in Congress, the emphasis is on cutting expenses, the EU's willingness to bail out its most debt-ridden members is faltering and everywhere central banks are taking measures to restrict loans, to defend their hoard.

At the same time, the proletariat, the population that has only its labour power to sell in order to survive, neither in China nor elsewhere is in the mood to sacrifice itself and is discovering new ways to fight, to communicate, to resist.

A collision is inevitable. As Bette Davis said: 'Fasten your seatbelts. It's going be a bumpy ride.'

Sander <sander_abroad@hotmail.com> is a New York based journalist who uses this pen-name for articles he couldn't possibly get published in the mainstream press

FOOTNOTES

1 See, amongst other texts, 'Value Creation and the Crisis of Capital', *Internationalist Perspective* 49, Spring/Summer 2008, http://internationalist-perspective.org/PI/pi-archives/pi_49_value.html

2 But the US produces almost as much, 19.4 percent, with almost a tenth of the labour force: there are only 11.5 million industrial workers left in the US. So the productivity gap remains considerable. See *Financial Times*, 13 March 2011.

3 See 'Capitalism, Technology and the Environment', *Internationalist Perspective* 50, Winter 2009, http://internationalist-perspective.org/PI/pi-archives/pi_50_environment.html and Sander, 'Artificial Scarcity in a World of Overproduction: An Escape that Isn't', Mute Vol2 #16, http://www.metamute.org/en/content/artificial_scarcity_in_a_world_of_overproduction_an_escape_that_isn_t

4 Of the 15 largest economies, China is second in income inequality, after Brazil. It is a sad irony that the greatest inequalities of the world are managed by the 'Communist Party' and the 'Party of Workers'.

5 Keith Bradsher, 'As China's Workers Get a Raise, Companies Fret', *The New York Times*, 31 May 2011, http://www.nytimes.com/2011/06/01/business/global/01wages.html

6 The last one is in part instigated by the State. It inflates the bubble because it profits from it: 'Through taxes, fees and property sales, local governments are raising more and more at the expense of the household sector's income and purchasing power. Local governments are essentially on a treadmill of raising more and more revenue to fund fixed investment. So it needs land prices to rise higher and higher, resulting in a massive and nationwide property bubble.' Andy Xie, 'Rebalancing cannot wait', *Caixin Weekly*, 11 March 2011, http://english.caing.com/2011-03-11/100235531.html

7 In 1998 the higher education institutions delivered 830,000 graduates, in 2009 6 million. Between 1982 and 2005, the number of graduates rose sevenfold while the number of white collar jobs rose from 7 to 13 percent

PROUD TO BE FLESH:

A MUTE MAGAZINE ANTHOLOGY OF CULTURAL POLITICS AFTER THE NET

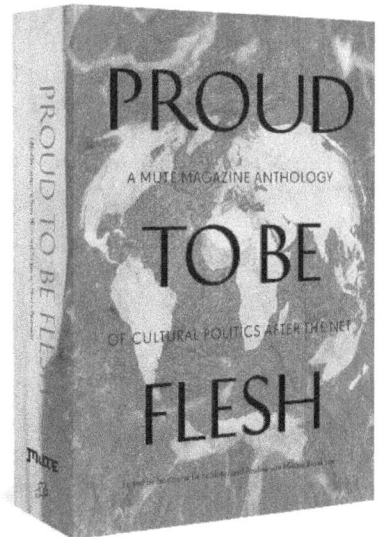

Edited by Josephine Berry Slater and Pauline van Mourik Broekman with Michael Corris, Anthony Iles, Benedict Seymour, and Simon Worthington

Compiling 15 years of *Mute* content, *Proud to be Flesh* offers 624 pages of the magazine's best writing, including interviews, essays, polemics and more. Divided into nine chronologically arranged chapters treating key themes associated with the 'digital revolution', *Proud to be Flesh* provides a unique history of a turbulent era and an excellent teaching tool

Hardcover **£44.99** Softcover **£24.99**

Proud to be Flesh can be purchased at all good bookshops, or previewed and ordered online at **metamute.org/proudtobeflesh**

To make a credit card order call **+44 (0)20 3287 9005**

For further enquiries, contact Howard Slater on <**howard@metamute.org**>

Published by Mute Publishing in association with Autonomedia

Softcover ISBN 978-1-906496-28-9
Hardcover ISBN 978-1-906496-27-2

'This collection of articles from the many incarnations of the *Mute* project is a great read, and a summation of that remarkable period of recent British history running from 1994 to 2009.'
-James Heartfield, *Spiked Review of Books*

'Essential reading for anyone interested in the ways in which evolving technology and business practices transform our culture - and how we might oppose such influences.'
-David Barrett, *Art Monthly*

'At a time when recent advances in digital technologies are still considered innovative yet remain an unexplored field for many of us, *Mute* can already claim scholarship in this area. I think *Proud to be Flesh* is an invaluable reference tool for researchers and it should be on the desks of all digital media curators and educationalists.'
-Nayia Yiakoumaki, Archive Curator, Whitechapel Gallery

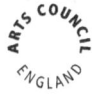 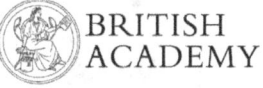

Supported by Arts Council England and The British Academy

BACK CATALOGUE:
OPENMUTE PRESS AND MUTE BOOKS

Since 2005, OpenMute Press has been helping artists, writers, and other independent producers bring their book ideas to fruition using Print On Demand, Short run Press (which is also used to make our magazine) and, more recently, eBooks. Our back-catalogue also includes the first titles from Mute Books and Mute Publishing's anthology, *Proud To Be Flesh*. In addition to all the usual online retailers, these diverse publications are made available through Metamute.

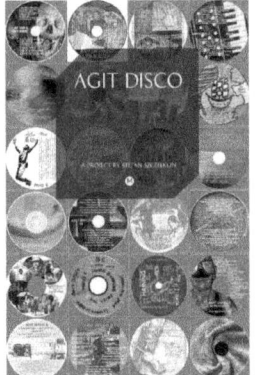

Stefan Szczelkun and Anthony Iles (Eds.)

Agit Disco

(Jan 2012)

Agit Disco collects the playlists of its 23 writers to tell the story of how music has politically influenced and inspired them. The book provides a multi-genre survey of political musics, from a wide range of viewpoints, that goes beyond protest songs into the darker hinterlands of musical meaning. Each playlist is annotated and illustrated. The collection grew organically with an exchange of homemade CDs and images. These images, with their DIY graphics, are used to give the playlists a visual materiality.

ISBN – 978-1-906496-51-7 / £11.99

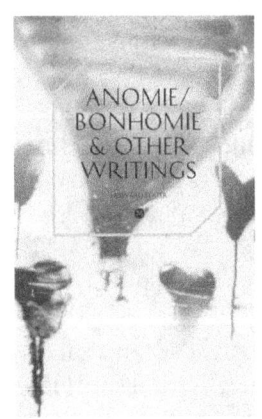

Howard Slater

Anomie/Bonhomie & Other Writings

(Jan 2012)

In this collection of writings, Howard Slater improvises around what Walter Benjamin could have meant by the phrase 'affective classes'. This 'messianic shard' and its possible implications leads Slater to develop a therapeutic micro-politics by way of a mourning for the Workers' Movement and a grappling with the 'becomings of capital'.

ISBN – 978-1-906496-72-2 / £9.99

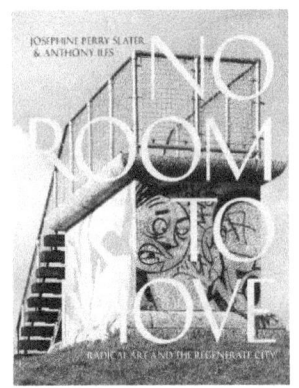

Josephine Berry Slater and Anthony Iles (Eds.)

No Room to Move: Radical Art and the Regenerate City

(Sep 2010)

As the Creative City model for urban regeneration founders, Anthony Iles and Josephine Berry Slater take stock of an era of highly instrumentalised public art making. Focusing on artists and consultants who have engaged critically with the exclusionary politics of urban regeneration, their analysis locates such practice within a schematic history of urban development's neoliberal mode. Featuring projects and interviews with Alberto Duman, Freee, Nils Norman, Laura Oldfield Ford and Roman Vasseur.

ISBN – 9781906496425 / £14.95

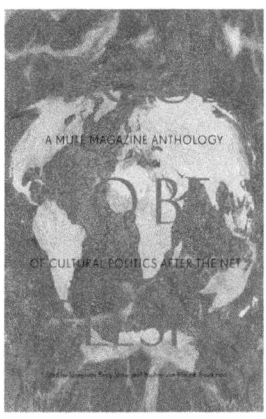

Josephine Berry Slater and Pauline van Mourik Broekman (Eds.) with
Michael Corris, Anthony Iles, Benedict Seymour, Simon Worthington

Proud to be Flesh: A Mute Magazine Anthology of Cultural Politics After the Net

(Nov 2009)

Proud To Be Flesh offers an expansive collection of some of *Mute*'s finest articles and is thematically organised around key contemporary issues: Direct Democracy and its Demons; Net Art to Conceptual Art and Back; I, Cyborg: Reinventing the Human; of Commoners and Criminals; Organising Horizontally; Art and/against Business; Under the Net - the City and the Camp; Class and Immaterial Labour; The Open Work.

Softback
ISBN – 9781906496289 / £24.95

Hardback
ISBN – 9781906496272 / £44.99

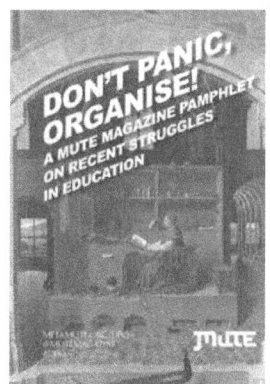

Don't Panic, Organise!: A Mute Magazine Pamphlet on Recent Struggles in Education

(Dec 2010)

From the introduction: 'They should be understood as part of the more gradual process of what George Caffentzis, in his analysis of the international situation, calls the "breakdown of the edu-deal"; the inability for capital, and therefore the state, to pay for the costs of producing a well educated workforce or to guarantee that investment in education will result in a more vigorous economy and increased living standards for those with qualifications.'

ISBN – 9781906496548 / £2.99

eBook
ISBN – 9781906496555 / Free

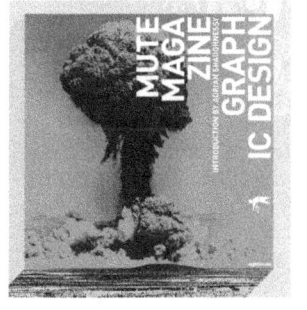

Damian Jaques, Pauline van Mourik Broekman, Adrian Shaughnessy and Simon Worthington (Eds.)

Mute Magazine Graphic Design

(May 2008)

In the early 1990s, long before the Internet became an integral part of life, a handful of pioneering magazines took it upon themselves to imagine the Internet into existence using fiction, interviews, speculative theory and experimental graphic design. Founded by artists Simon Worthington and Pauline van Mourik Broekman, London based Mute occupied a central position. *Mute Magazine Graphic Design* presents and contextualises its graphic output. [Published by Eight Books]

ISBN – 9780955432224 / £19.95

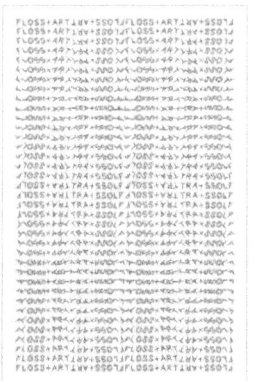

Aymeric Mansoux and Marloes de Valk (Eds.)

FLOSS + Art

(Aug 2008)

FLOSS+Art reflects critically on the growing relationship between Free Software ideology, open content and digital art. It provides a view onto the social, political and economic myths and realities linked to this phenomenon. Contributors: Fabianne Balvedi, Florian Cramer, Sher Doruff, Nancy Mauro Flude, Olga Goriunova, Dave Griffiths, Ross Harley, Martin Howse, Shahee Ilyas, Ricardo Lafuente, Ivan Monroy Lopez, Thor Magnusson, Alex McLean, Rob Myers, Alejandra Maria Perez Nuñez, Eleonora Oreggia, oRx-qX, Julien Ottavi, Michael van Schaik, Femke Snelting, Pedro Soler, Hans Christoph Steiner, Prodromos Tsiavos, Simon Yuill.

ISBN – 9781906496180 / £18.50

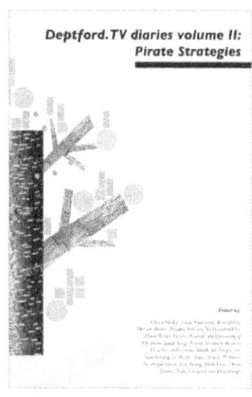

Deptford.TV

Deptford.tv Diaries II – Pirate Strategies

(Apr 2008)

This reader problematises the notion of 'tactical media' – calling for a more strategic approach. Contributors: Adnan Hadzi, Jonas Andersson, Ben Gidley, Duncan Reekie, Brianne Selman, Neil Gordon-Orr, Alison Rooke, Gesche Wuerfel, the University of Openness, Jamie King, Armin Medosch, Rasmus Fleischer, andrea rota, Bitnik Mediengruppe, Sven Koenig Jo Walsh, Rufus Pollock, Platoniq, The People Speak, Zoe Young, Mick Fuzz, Denis Jaromil Rojo, Lennaart van Oldenborgh.

ISBN – 9781906496111 / £5

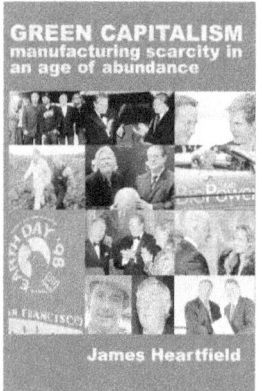

James Heartfield

Green Capitalism: Manufacturing Scarcity in an Age of Abundance

(Feb 2008)

A polemic against 'Green Capitalism', which James Heartfield accuses of profiteering from climate change and other environmental scares. Green capitalists like Zac Goldsmith and Al Gore are manufacturing scarcity to boost prices. The technological revolution has removed scarcity from most of our lives, but the green capitalists are trying to re-invent it.

ISBN – 9781906496104 / £7.50

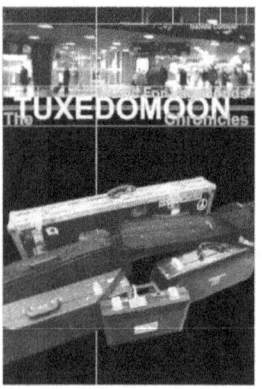

Isabelle Corbisier

Music for Vagabonds: the Tuxedomoon Chronicles

(Jan 2008)

Tuxedomoon is a group of musicians and performers that was formed in San Francisco in 1977. Their identity is as elusive as their geographical location. Tuxedomoon have attracted followers and gained cult status, never ceasing their quest for a permanently elusive and lost 'home' – some other America or the quaint Europe of their fantasies. From 2001 onwards, the author of this book found herself sucked into Tuxedomoon's spiral of vagrancy and travelled the world to meet the actors in this ongoing 30-year-old story.

ISBN – 9781906496081 / £19

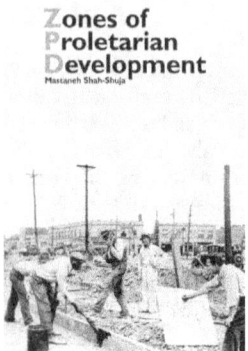

Mastaneh Shah-Shuja

Zones of Proletarian Development

(Jan 2008)

Zones of Proletarian Development is an attempt to theorise the anti-capitalist movement from a neo-Vygotskian perspective. Using Marx, Vygotsky, Bakhtin and Activity Theory, it analyses a series of proletarian activities including recent May Day celebrations in London, carnivalesque football riots in Iran, the anti-poll-tax rebellion and the anti-war movement. It concludes by looking at past and current proletarian organisations and makes a number of proposals for future modes of organising conducive to radical consciousness and autonomous activity.

ISBN – 9781906496067 / £15

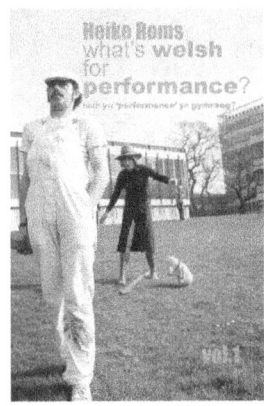

Heike Roms
What's Welsh for Performance?
(Dec 2007)

For more than forty years artists have been creating performances, happenings and other time-based art in Wales, yet their work remains largely confined to half-remembered anecdotes, rumours and hearsay. *What's Welsh for Performance?* tries to uncover Wales' hidden history of performance in conversations with key artists who have shaped this history since 1968. With: Shirley Cameron, Ivor Davies, Anthony Howell, John Chris Jones, Timothy Emlyn Jones, Andrew Knight, Roland Miller. Dr. Heike Roms is lecturer in Performance Studies at Aberystwyth University, Wales.

ISBN – 9780955392726 / £10

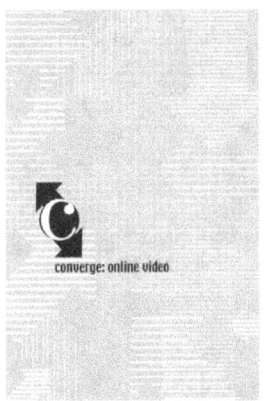

Converge – Online Video
(Oct 2007)

Using media as a means of working with, and empowering marginalised people in their communities is a practice that has emerged strongly in recent years, nurtured by the extraordinary growth of digital media and the Web. These developments have enabled a participatory culture – particularly online – in which young people are now more able to represent themselves and their concerns. This book offers first hand accounts of work across and beyond Inclusion Through Media, alongside critical analysis of many of the processes involved, and the policy issues it raises. The book includes an accompanying DVD.

ISBN – 9781906496005 / £9.99

E. Karaba (Ed.)
Feedback 4: Ideas that Inform, Construct and Concern the Production of Exhibitions and Events
(Sep 2007)

FeedBack 4 focuses on participatory art events and examines them from the point of view of the artist, the curator and the participant. It brings together contributions from many authors interested in curatorial debates.

ISBN – 9780955479687 / £6.50

Steven Dickie
New Chronica Dublin
(Jul 2007)

New Chronica Dublin is a graphic novel and musical album download, based on a collection of contemporary folklore from the Irish capital. The dilution of identity and the inevitable displacement of population which the rebranding of economic and social areas brings, may result in this being the last opportunity to witness the community's galvanised common identity as it negotiates a position within the new Dublin.

ISBN – 9780955479663 / £10

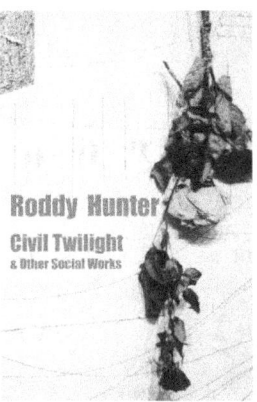

Roddy Hunter

Civil Twilight & Other Social Works

(Feb 2007)

Civil Twilight & Other Social Works explores the performance artwork of provocative Scots artist Roddy Hunter. Through the artist's own texts and archival documentation, Hunter introduces us to his methodology of research into the idea of urban civic centres as places where collective identity is formed.

ISBN – 9780955392719 / £10

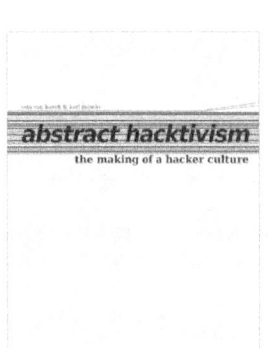

Otto von Busch & Karl Palmås

Abstract Hacktivism: The Making of a Hacker Culture

(Dec 2006)

In recent years, designers, activists and business people have started to navigate their social worlds on the basis of concepts derived from the world of computers and new media technologies. According to Otto von Busch and Karl Palmås, this represents a fundamental cultural shift. In the nineteenth century, the motor replaced the clockwork as the universal model of knowledge; new media technologies are currently replacing the motor as the dominant 'conceptual technology' of contemporary social thought.

ISBN – 0955479622 / £5

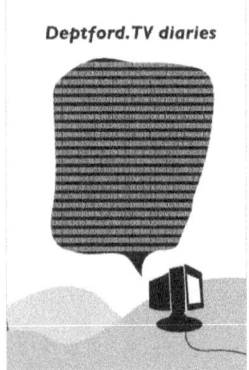

Deptford TV

Deptford.tv Diaries

(Dec 2006)

Deptford.TV is an audio-visual documentation of the regeneration process of Deptford (south-east London) in collaboration with SPC.org media lab, Bitnik.org, Boundless.coop, Liquid Culture and Goldsmiths College. Contributors: Adnan Hadzi, Maria X, Heidi Seetzen, James Stevens, Erol Ziya, Bitnik media collective, Andrea Pozzi, Andrea Rota and Jonas Andersson, alongside selected public license texts from Hakim Bey, Jaromil and Guy Debord.

ISBN – 0955479606 / £5

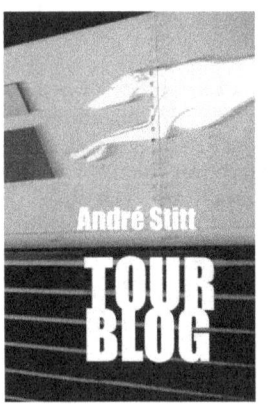

André Stitt

Tour Blog 2006: On Tour with Panacea Society USA

(Sep 2006)

No Sex, No Drugs, No Rock 'n' Roll. Just freak-beat, psyche-garage, prog-techno and artyness. Who do they think they are, these performance artists who want to be the new Golden Gods of Art-Rock? Find out as you join The Panacea Society on their 2006 North American tour. When asked by one of his students what he would be doing during the Easter break, performance artist André Stitt decided to write a tour diary.

ISBN – 0955392705 / £10

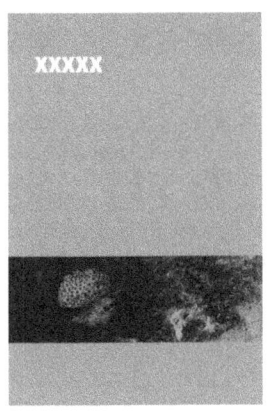

xxxxx

(Sep 2006)

[The] xxxxx [reader] proposes a radical, new space for artistic exploration, with essential contributions from a diverse range of artists, theorists, and scientists. Combining intense background material, code listings, screenshots, new translations, [the] xxxxx [reader] functions as both guide and manifesto for a thought movement which is radically opposed to entropic contemporary economies. xxxxx traces a clear line across eccentric and wide ranging texts under the rubric of life coding which can well be contrasted with the death drive of cynical economy with roots in rationalism and enlightenment thought.

ISBN – 0955066441 / £15

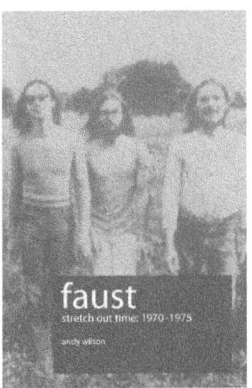

Andy Wilson

Faust: Stretch Out Time 1970-1975

(Jun 2007)

In 1970 Polydor Records funded an unusual experiment. They gave some unknown German musicians a retreat in the countryside near Hamburg, equipped it with a studio and their best engineer, then left them free to do as they liked. This is the story of Faust and the music they made between 1970 and 1975, music which continues to inspire and confound listeners to this day.

ISBN – 095506645X / £10.99

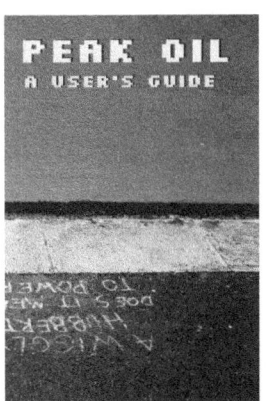

Popex

Peak Oil: A User's Guide

(July 2006)

This book is composed of two parts. The first part is the manual for the Peak Oil Olympics, a happening which took place on the 1st and 2nd July 2006 in Bristol, UK. This has been updated with documentation from the event, and expanded by the second part which is a look at Peak Oil from a more critical angle incorporating texts by George Caffentzis and Iain Boal which provide a more systematic analysis of what Peak Oil might mean.

ISBN – 0955066468 / £4

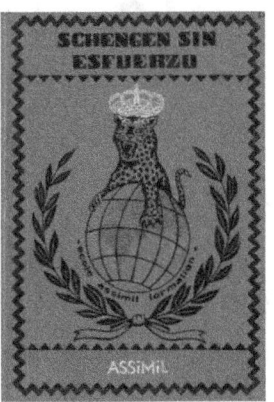

Vahida Ramujkic

Schengen with Ease

(Feb 2007)

'Extra-comunitarios', or citizens of non-European countries, have the 'extra' bureaucratic task of changing their status, to one that will allow them to move and work 'freely' within the European Union. The length and complexity of this process can vary depending on the type of 'extra-comunitario' in question. Almost everyone agrees that bureaucracy is the most boring thing on the world. *Schengen with Ease* is a compilation of material from a variety of official and non-official sources, brought together to explain how daily practices are affected by the application of the EU Foreign Legislation and the Schengen Agreement in the territory of the European Union.

ISBN – 0955066484 / £8.29

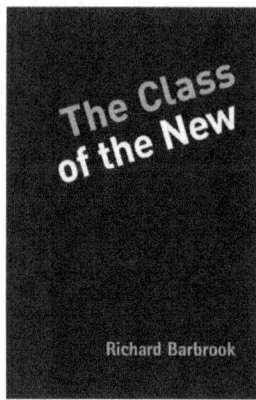

Richard Barbrook

Class of the New

(May 2006)

Netizens, elancers, cognitarians, swarm-capitalists, hackers, produsumers, knowledge workers, pro-ams... these are just a few of the monikers that have been applied to the new social class emerging from the networked workplace. In this short book, Richard Barbrook presents a collection of quotations from authors who in different ways attempt to identify an innovative element within society - 'the class of the new'. Announcing a new economic and social paradigm, this class constitutes a 'social prophecy' of the shape of work to come.

ISBN –0955066476 / £4

NODE.London

A NODE.London Reader - A Survey of Media Arts, Technologies and Politics

(Feb 2006)

The NODE.London reader, *Media Mutandis*, projects a critical context around the Season of Media Arts in London March 2006 and provides another discursive dimension to the events of October 2005's Open Season. It engages debates in FLOSS (Free/Libre and Open Source Software), media arts and activism, collaborative practices and the political economy of cultural production in the present day. Contributions from Sabeth Buchmann, Toni Prug, Armin Medosch, Simon Yuill, Chad McCail, Critical Art Ensemble, Jo Walsh, Richard Barbrook, Michael Corris, Harwood, Kate Rich, Agnese Trocchi, Matthew Fuller, Rasmus Fleischer and Palle Torsson, Brett Neilson and Ned Rossiter, Matteo Pasquinelli and Francis McKee.

ISBN – 0955243505 / £5

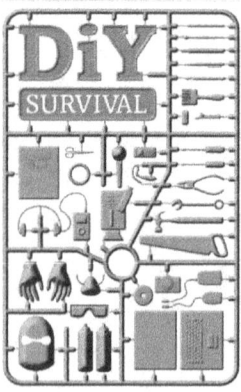

C6

DiY Survival

(Oct 2005)

DiY Survival - There is no subculture, only subversion. DiY or do-it-yourself survival is a collection of essays, tips and case studies collated from an online call for participation by the maverick art group C6. The eclectic mix presented within these pages shows the breadth and diversity of art/activism practice today. Whether that is creating wireless networks, pissing on national monuments or building cardboard friends, it is certain that these submissions show that practitioners are taking their work to new spaces and audiences, redefining, through the engagement with the community, what we have considered to be 'art'.

ISBN –0955066492 / £4.99

WWW.METAMUTE.ORG/SHOP

Please get in touch with us on **mute@metamute.org** with any queries about producing your book through OpenMute.

Further information and pricing is also available at **www.openmute.org**

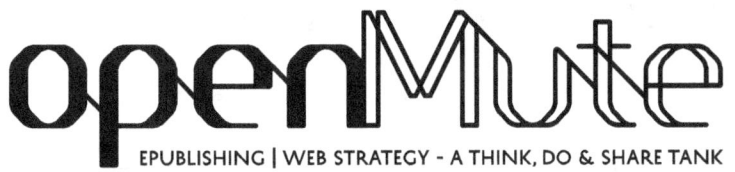

openMute

EPUBLISHING | WEB STRATEGY - A THINK, DO & SHARE TANK

EPUBLISHING

POD BOOK PRINTING

We provide a service to consult, set-up and manage a book title or library into digital print-on-demand book printing, from mass produced paperbacks to bespoke handmade artists' books

EBOOKS/ KINDLE AND IPHONE BOOKS

OpenMute can create a variety of eReader book formats and place them into the relevant distribution and sales platforms, providing you with reports, user stats and sales payments

BOOK DESIGN, PRODUCTION AND EDITORIAL

The OpenMute team have years of experience in book design and layout, as well as production management and editorial work, proofing, style guides, indexing etc.

WEB

DIGITAL STRATEGY

We can provide consultancy and workshops on how to develop your organisation online, communicate your message, grow audiences and increase revenues. Covering areas like social networking, basic infrastructures and audience development & measurement

DRUPAL SITE DESIGN AND BUILD

The Drupal CMS is versatile and easy to use, with a large Open Source community supporting it. This means that new Web 2.0 technologies can be quickly integrated

SOCIAL NETWORKING AND WEB 2.0

We can recommend and integrate social networking and Web 2.0 tools that offer you low cost professional functionality. To name a few: Facebook, Twitter, Amazon store, MailChimp, Vimeo, Mobile Roadie, Sound Cloud and LibraryThing

CONTACT:OPENMUTE
EMAIL: SIMON@METAMUTE.ORG
SKYPE: MUTE.LONDON
TEL: +44 (0)20 3287 9005

MUTE BOOKS

Mute Books is the imprint series of *Mute* magazine. It will specialise in cultural politics and give deserved space to the many distinctive voices and practitioners that the magazine has hosted since its inception in 1994

NEW TITLES

ANOMIE/BONHOMIE & OTHER WRITINGS

BY HOWARD SLATER

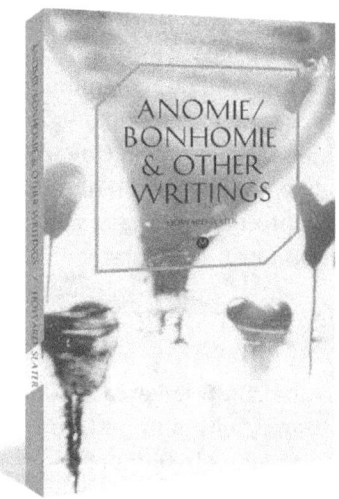

AGIT DISCO

A PROJECT BY STEFAN SZCZELKUN

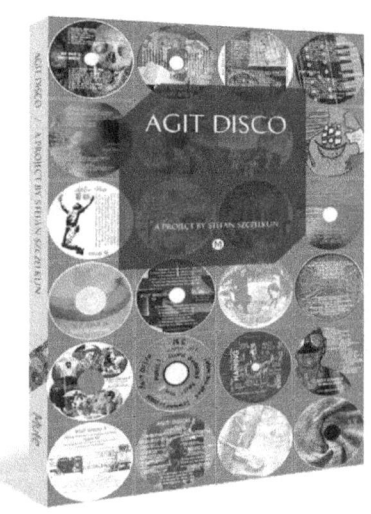

In this collection of writings, Howard Slater improvises around what Watter Benjamin could have meant by the phrase 'affective classes'. This 'messianic shard' and its possible implications leads Slater to develop a therapeutic micro-politics by way of a mourning for the Workers' Movement and a grappling with the 'becomings of capital'.

Publication: 2012 Price: **£9.99**
ISBN 978-1-906496-72-2

Edited by Stefan Szczelkun and Anthony Iles

Taking its cue from mix-tape culture the Agit-Disco project, initiated in 2007, now transmutes from an on line project and CD-R distribution process into book form. Agit-Disco brings together the music selections of its invited participants and covers a range of genres and styles that are offered, here, as a collective response to the remit: politics and music minus the propaganda.

Published: 2012 Price: **£11.99**
ISBN 978-1-906496-51-7

For further enquiries, contact Caroline Heron on <caroline@metamute.org>

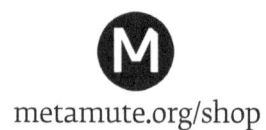

metamute.org/shop

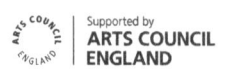

Supported by
**ARTS COUNCIL
ENGLAND**

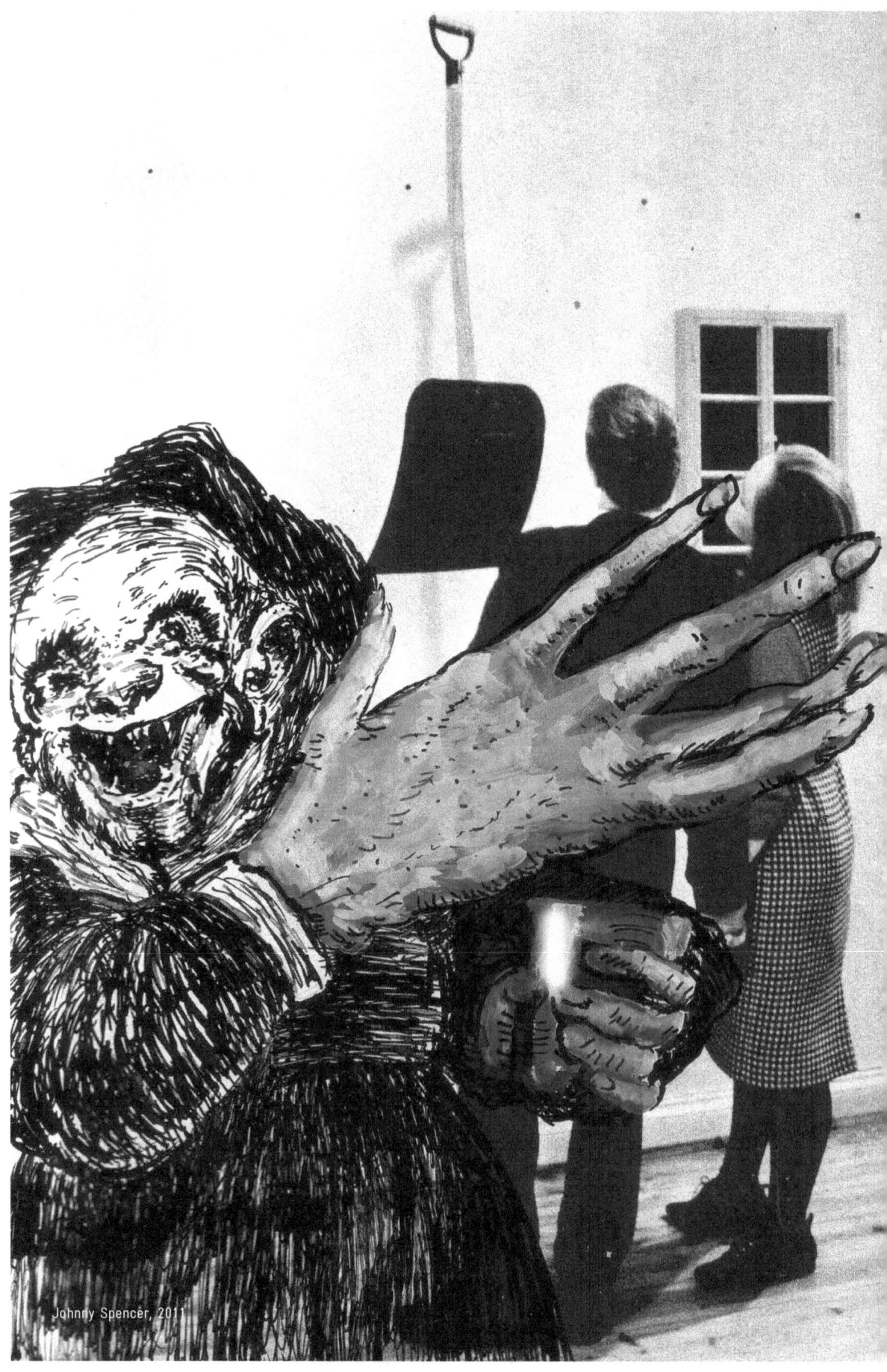

Johnny Spencer, 2011

www.ingramcontent.com/pod-product-compliance
Lightning Source LLC
Chambersburg PA
CBHW082007230526
45468CB00023B/2685

* 9 7 8 1 9 0 6 4 9 6 7 9 1 *